MULTIPLICATION OBJECTS IN MONOIDAL CATEGORIES

MULTIPLICATION OBJECTS IN MONOIDAL CATEGORIES

JOSÉ ESCORIZA LÓPEZ

NOVA SCIENCE PUBLISHERS, INC.
Huntington, NY

Editorial Production: Susan Boriotti
Office Manager: Annette Hellinger
Graphics: Frank Grucci, Drew Kane and Jennifer Lucas
Information Editor: Tatiana Shohov
Book Production: LaToya Clay, Donna Dennis, Cathy DeGregory and Lynette Van Helden
Circulation: Anna Cruz, Ron Hedges and Andre Tillman

Library of Congress Cataloging-in-Publication Data

López, José Escoriza.
Multiplication objects in monoidal categories / José Escoriza López
p. cm.
Includes bibliographical references and index.
ISBN 1-56072-823-X.
1. Categories (Mathematics) 2. Torsion theory (Algebra) I. Title.

QA169 .L67 2000
511.3—dc21 00-041148

227 Main Street, Suite 100
Huntington, New York 11743
Tele. 631-424-6682 Fax 631-424-4666
E Mail Novascil@aol.com

Printed in the United States of America

Contents

Preface 5

Introduction 7

1 Torsion theories and monoidal categories 25

1.1 Commutative torsion theories 26

1.2 Monoidal categories 40

2 Multiplication objects in monoidal categories 47

2.1 Properties of multiplication objects 48

2.2 Commutative Grothendieck categories 51

2.3 Some examples of multiplication objects 66

2.4 Endomorphisms of multiplication objects 72

3 Multiplication objects in the category of relative injective relative torsionfree modules 79

3.1 Alternative definitions 80

3.1.1 Multiplication elements in semimodules 80

3.1.2 Multiplication modules relative to a torsion theory . . 84

3.2 Localization with respect to torsion theories 95

3.3 Construction of relative multiplication modules 104

3.4 Relative distributive modules 118

3.5 Relative multiplication ideals and rings 128

4 Multiplication objects in gr-R **145**

4.1 Properties of multiplication graded modules 146

4.2 Graded localization 150

4.3 Distributive graded modules 160

4.4 Multiplication graded rings 164

4.5 Graded rings and multiplication rings 170

Bibliography **179**

Index **192**

Preface

The main aim of the book is to study the concept of multiplication object from a categorical point of view, namely, in the setting of monoidal categories. They are responsible for the narrow relationship between quantum groups and knot theory. At the same time, the book makes accessible to any mathematician a topic, multiplication modules and rings, whose results have been scattered in the literature up to date. Now, they are organized and exposed in a categorical framework, by using functorial techniques.

Multiplication modules and rings are framed inside commutative algebra, which is a basis for number theory and algebraic geometry. They include families of rings very important in ideal arithmetic such as regular von Neumann rings, Dedekind domains, hereditary rings or special primary rings. In the case relative, i. e., multiplication modules and rings with respect to a hereditary torsion theory, the most significant example is that of Krull domains (with respect to the classical torsion theory). As a consequence, we have an adequate setting to consider divisorial properties. As for the graded concept, it is possible to deep in the study of arithmetically graded rings such as generalized Rees rings, graded Dedekind domains, twisted group rings, etc. The book points out some different possibilities to deal with the topic, for example, semiring theory, lattice theory, comodule theory, etc.

The only prerequisites are a basic knowledge of results from commutative algebra. Every concept is tried to be accessible to any nonspecialist. Therefore, the book can be used for a postgraduate course of multiplicative ideal theory. It has a special interest for specialist in arithmetical (graded) rings, torsion theories or monoidal categories.

Introduction

The first cornerstone which the book is based on is the concept of multiplication module over a commutative ring. This notion is placed in the context of the multiplicative ideal theory that is a part of commutative algebra, which constitutes a basis for the study of number theory and algebraic geometry.

W. Krull introduced the concept of multiplication ring in [68]. A commutative ring R is multiplication if given two ideals $A \subseteq B$, there exists an ideal C of R such that $A = B.C$. This family includes von Neumann regular rings, hereditary rings and special primary rings (or, more briefly, SPIR's). A *special primary ring* is a local ring in which every proper ideal is a power of its maximal ideal.

In [80], J. L. Mott gave several characterizations of multiplication rings. He proved that R is multiplication if and only if every prime ideal is multiplication, i. e., given ideals $A \subseteq P$, P being prime, there exists $C \leq R$ such that $A = P.C$. Now, this result is known as Mott's Theorem. In a joint work with R. Gilmer (cf. [50]), he proved that every multiplication ring verifies that its ideals with prime radical are primary. He generalized the above work for rings without identity element. By using their techniques, they showed that indecomposable multiplication rings are Dedekind domains and SPIR's (this result was proved before by Mori, who showed that in the noetherian case, multiplication rings are just finite direct sums of Dedekind domains and SPIR's). In the general case, it is well-known that multiplication rings are subrings of a direct product of Dedekind domains and SPIR's. However, there are examples of multiplication rings which are not a product of Dedekind domains and SPIR's. Let R be the set of elements consisting of series of elements of $\mathbb{Z}_2$ which are constant from a term further on. R is a ring where every element is idempotent and therefore, is multiplication. R is a subring of the product with index set $\mathbb{N}$ of copies of $\mathbb{Z}_2$, R'. But $R \neq R'$ as in R' there is an uncountable number of elements and in R, a countable amount. In [81], Mott proved that R is a noetherian multiplication ring if and only if R is multiplication containing only many finitely minimal prime ideals. It is also proved that R is a noetherian multiplication ring if and only if every non-zero ideal is a product of prime ideals, i.e. R is a *ZPI-ring* (Zerlegung Primideale) (cf. [69, Theorem 9.10]). The Krull dimension of multiplication rings is proved to be smaller or equal to one.

W. W. Smith (cf. [104]) showed that every hereditary ring is multiplication by using the dual basis theorem in order to prove that every projective ideal is a multiplication ideal. He proved that every finitely generated multiplication ideal of a local ring (a ring with a unique maximal ideal) is principal.

M. Griffin studied multiplication rings by means of their total quotient rings (cf. [57]). He proved that a multiplication ring is semihereditary if and only if it has no nilpotent elements and that every finitely generated ideal in a multiplication ring is generated by two elements.

D. D. Anderson studied multiplications ideals and improved the result by W. W. Smith by showing that every multiplication ideal (not necessarily finitely generated) in a local ring is principal. Mott's Theorem is also slightly improved by showing that R is multiplication if and only if its localizations in maximal ideals are multiplication rings. By using these results, he proved that the polynomial ring $R[X]$ is multiplication if and only if R is a finite direct sum of fields (cf. [8] and [10]).

In [52], R. W. Gilmer collects investigations by L.G. Chouinard, B. R. Hardy and T. S. Shores about when a commutative semigroup ring $R[G]$ is a multiplication ring. For example, if G is a torsion-free group, $R[G]$ is multiplication if and only if R is a finite direct sum of fields and $G \cong \mathbb{Z}$.

Multiplication rings have been related to other classes or rings in the literature. In [69], they appear in a chapter devoted a generalizations of Dedekind domains together with almost Dedekind domains (rings which are locally Dedekind domains), ZPI-rings and almost multiplication rings (also studied by H.S. Butts and R.C. Phillips in [20]). In [51], it is proved that ZPI-rings and multiplication rings coincide in the noetherian case. There is a narrow relationship between arithmetical rings and multiplication rings. A ring R is *arithmetical* if the lattice of ideals has distributive property of the sum with respect to the intersection. They were introduced by L. Fuchs in 1949 (cf. [45]). After that, Jensen related them to multiplication rings, by proving that R is arithmetical (or distributive) if and only if every finitely generated ideal is a multiplication ideal, i. e., it verifies the condition stated in the definition of multiplication ring for any finitely generated ideal. As for domains, arithmetical domains are exactly Prüfer domains or semihereditary domains. P. J. McCarthy studied arithmetical rings from the point of view of multiplicative lattices. V. Camillo proved that the polynomial ring $R[X]$ is distributive if and only if R is von Neumann regular if and only if $R[X]$ is semihereditary. B. R. Hardy and T. S. Shores studied arithmetical semigroup rings.

V. Erdoğdu studied regular multiplication rings in [35]. A ring is *regular multiplication* if every regular ideal is multiplication. They are characterized as rings R verifying that R/I is a finite direct sum of SPIR's for every regular ideal I. An example of regular multiplication ring which is not multiplication is the ring $R = K[x^2, x^3]/(x^4)$.

The first definition of multiplication module is due to F. Mehdi (cf. [75]). Since this definition is different from the one we shall use, we say that an R-module M is *strongly multiplication* if every submodule is an multiplication R-module. Obviously, a ring R is a multiplication ring if and only if it is strongly multiplication as an R-module. Mehdi proved that the only free multiplication modules are those of rank one and he answered the questions of when the direct sum or the tensor product of strongly multiplication modules are strongly multiplication modules as well. It is proved that if an integral domain R has a faithful multiplication module, then R is a Dedekind domain and the lattices of ideals of R and submodules of M are isomorphic.

It was A. Barnard who gave the definition of multiplication more spread and modern in 1981 (cf. [13]). A right R-module M is *multiplication* if for every submodule N, there exists an ideal I such that $N = MI$. To exemplify this theory we can think of cyclic modules, projective ideals or finitely generated submodules of distributive modules (they will be defined further on). He proved that a module is distributive if and only if every finitely generated submodule is multiplication and characterized multiplication modules locally. Finitely generated multiplication modules are modules that locally are cyclic and every artinian finitely generated multiplication module is cyclic.

In [10], D.D. Anderson characterized multiplication ideals by associating to every ideal I the ideal $\theta(I) = \sum_{x \in I}(xR : I)$, where $(xR : I) = \{r \in R; Ir \subseteq xR\}$. He showed that I is multiplication if and only if $I_P = 0$ for every maximal ideal P containing $\theta(I)$. It is also proved that the concept of multiplication ideal is equivalent to that of *meet-principal ideal*, that is, an ideal I such that $AI \cap B = A \cap (B : I)$ for any ideals A, B of R.

In 1986, A.G. Naoum in a joint work with M. A. K. Hasan proved that a module M that is finitely generated by $m_1, \cdots, m_n$ is multiplication if and only if $\sum_{i=1}^{i=n}(m_i R : M_i) = R$, M_i being the submodule generated by all m_j's except m_i. He also studied the sum of finitely generated multiplication modules and after that, the relationship with projective modules, endomorphisms rings, the dual of a multiplication module and homomorphisms between a finitely generated multiplication module M and any multiplication module N (cf. [82], [83], [84], [85] and [86]). He proved that the endomorphism ring of a finitely generated multiplication R-module M is isomorphic to the commutative ring $R/ann(M)$, where $ann(M)$ stands for the annihilator of M in R. Moreover, if M is projective, finitely generated and *endocommutative* (with a commutative endomorphism ring), then M is multiplication. Let D be the double annihilator of the R-module M. If M is finitely generated and multiplication and D is finitely generated, then the dual module, M^* is multiplication if and only if D is a multiplication ideal of R. With W.K.H. Aubaidy, he proved that if M is a faithful multiplication R-module and S

is its endomorphism ring, then the following are equivalent: M is noetherian (artinian) as an R-module, R is noetherian (artinian), S is noetherian (artinian), M is a noetherian (artinian) S-module (cf. [87]).

In [31], working jointly with El-Bast, Patrick F. Smith investigated when multiplication modules are cyclic or finitely generated. They obtained that if M is artinian, then M is multiplication if and only if it is cyclic. If M is noetherian, M is multiplication if and only if there exist ideals $A \subseteq B$ such that $M \cong B/A$ and B/A is an invertible ideal in R/A. As a result, every artinian multiplication module is noetherian. They also proved that if R have descendent chain condition (D.C.C.) on semiprime ideals (an ideal I is *semiprime* if the ring R/I has no nonzero nilpotent elements), then every multiplication R-module is finitely generated. In the same year, P. F. Smith (cf. [102]) investigated when the sum, intersection and direct sum of multiplication modules is multiplication as well, generalizing results achieved by Naoum. He proved that if M is a faithful finitely generated multiplication module, then a submodule N, is multiplication if and only if the ideal $(N : M)$ is multiplication. By following the ideas by Mott, El-Bast and P. F. Smith proved that if every prime submodule is multiplication, then the module is multiplication (N is a *prime submodule* of M if $ma \in N$ implies $m \in N$ or $Ma \subseteq N$). In a joint work with G. Low ([72]), he proved that a faithful multiplication module M is isomorphic to an ideal if and only if M^* is non torsion. Projective and multiplication modules are related in [103]. He defined a *finitely projective module* as a module M that verifies that for every finitely generated submodule N, there exist an integer n, $m_i \in M$ and morphisms $\theta_i : M \to R$ $(1 \leq i \leq n)$ such that $x = m_1\theta_1(x) + \cdots + m_n\theta_n(x)$ for any $x \in N$. He obtained that M is multiplication if and only if M is finitely projective as an $R/ann(M)$-module and its endomorphism ring, $End(M_R)$, is commutative. The endomorphism ring of an indecomposable multiplication module with descendent chain condition on multiplication submodules is proved to be local.

Since arithmetical rings and multiplication rings are narrowly related, it seems natural to research into the relationship between distributive modules and multiplication modules. T. Albu and C. Năstăsescu (cf. [2]) defined *distributive modules* as those modules having a distributive submodule lattice, i. e., with distributive property of the sum with respect to the intersection (or, equivalently, to the intersection with respect to the sum). They proved that being distributive is a local property for a module, i.e., the R-module M is distributive if and only if M_P is distributive as an R_P-module. It is studied when the tensor product and the homomorphism module between distributive modules are distributive. W. Stephenson (cf. [106]) gave several arithmetical characterizations of distributive modules. In [22], V. Camillo showed that M is distributive if and only if for every submodule N, M/N has square-free socle, i.e., at most one copy of each simple module. P. Vámos showed that artinian finitely generated distributive modules are cyclic. In [34], V. Erdoğdu studied multiplication modules which are distributive and he obtained that over an arithmetical ring, multiplication modules are distributive. When M is a finitely generated faithful R-module, it is distributive if and only if M is multiplication and R is arithmetical or, equivalently, M is projective with rank one and R is arithmetical. He studied when a direct sum of distributive submodules is distributive in [33].

There have been other families of modules sharing many properties of multiplication modules. For example, Mehdi defined a prime submodule of M as a submodule $N \neq M$ such that $A.N_1 \subseteq N$ and $N_1 \not\subseteq N$ implies $AM \subseteq N$ for any ideal A of R and any submodule N_1 of M. An R-module M is said to be a *weak multiplication module* if every prime submodule is multiplication. The $\mathbb{Z}$-module $\mathbb{Q}$ is weak multiplication but it is not multiplication (cf. [76]).

A.G. Naoum and P. A. Hummadi defined semi-multiplication modules. Given a submodule N of the R-module M, let R' be the endomorphism ring of M. We write $[N : M]'$ for $\{\varphi \in R'; \varphi(M) \subseteq N\}$ and $\theta'(M) = \sum_{a \in M}[(a) : M]'$. Then, M is called *semi-multiplication* if $\theta'(M) = R'$. Every finite generated projective module is semi-multiplication. In another work with F. H. Al-Alwan, Naoum introduced and studied *Dedekind modules*, i. e., modules whose non-zero submodules are invertible (cf. [88]).

Although it is well-known that associating to a ring R the category of right R-modules, Mod-R, we can deduce properties of R from those of Mod-R, this technique provides us some information on properties of rings which are invariant under Morita equivalences. We need another powerful technique, which will be very often used in this book, namely, torsion theory. Every ring has an associated torsion theory lattice, which is complete and has distributivity of intersection with respect to infinite unions (such a lattice is called a *frame*). From properties of the lattice or some elements of the lattice, we can determine properties of Mod-R and R. Torsion theories generalize the classical concepts of torsion of abelian groups and localization in prime ideals. In the beginning, torsion theories were defined over arbitrary abelian categories, being outstanding in this sense the works by

Gabriel, Dickson and Walker (cf. [30], [46] and [116]).

Torsion theories for commutative rings have been studied by many authors (cf. [15], [18], [19], [21] and [90]) and provide us functorial techniques which have been extremely useful. Between the rings characterized by means of torsion theories, we emphasize semicentered, seminoetherian, semiartinian and convenient rings (cf. [53, Themes 49,50, 51 and 52] and [96]). Related to torsion theories, we have used works by Chew on closure operations in quotient rings (cf. [26]) and by D. Kirby on closure operations on submodules (cf. [67]). Năstăsescu defined and studied arithmetical rings with respect to a semicentered torsion theory τ (he calls it "stable"). A. C. Benander works with projective concepts in torsion theories that will be developed in our setting. W. Brandal and E. Barbut investigated localization of torsion theories and they described all Gabriel topologies for valuation domains (cf. [18]). In the same way as many of the notions related to a a torsion theory τ, that of τ-multiplication module can be interpreted as a multiplication object in a certain quotient category of Mod-R. In this way, it appears the interest to study multiplication objects in more general categories than module categories.

A second group of tools which will be applied in this work consists of graded techniques. Graded rings appear in the study of completions of commutative rings and they are basic to study projective algebraic geometry, representation theory and differential operators rings. As for the theme of this work, it is necessary to point out works by F. Van Oystaeyen and his students on graded Dedekind rings, generalized Rees rings and other arithmetically graded rings, such as graded Krull rings (cf. [27], [28], [29], [70],

[112], [113] and [114]). Several results of group rings are generalized and it is answered when a certain property (being a Dedekind domain, being a principal ideal ring, etc.) holds for graded rings. In [113], Van Oystaeyen gave structure theorems for gr-Dedekind domains and gr-principal ideal domains. When working with $\mathbb{Z}$-graduations, he related the above classes of rings to the class of the part of degree zero. Generalized Rees rings were also related to gr-Dedekind domains. We have taken into account works by E. Jespers and F. Decruyenaere. In [27], they characterized Dedekind domains graded by a nontrivial torsionfree group. In [28] they continued with the study of divisibility properties of graded rings and they achieved structure theorems for graded principal ideal rings. In [29], they obtained partial characterizations of Prüfer domains when they are graded rings. For example, if R is a Prüfer domain strongly graded by $\mathbb{Z}$, then $R \cong K[x, x^{-1}]$, where K is a field. Almost Dedekind domains graded by a torsionfree abelian group were also completely determined.

Monoidal categories were introduced by Bénabou and Mac Lane in 1963 (cf. [17] and [73]). They appear by axiomatizing the properties of tensor product in the Hopf algebra representation category. They are responsible for the narrow relationship between Quantum Groups and Knot Theory, according to [66]. They constitute the adequate setting for Hopf algebra representations. Therefore, mathematicians and theoretical physicists have interested in them. Joyal and Street (cf. [64]) call them *tensor categories.* D. N. Yetter showed up their relationship with classical theory of knots and links, quantum integrable systems, statistical mechanics and quantum field theory (cf. [118]). With respect to this work, they constitute an ideal framework to link the concepts of multiplication module, multiplication module relative to a torsion theory, multiplication graded module and multiplication object in a commutative Grothendieck category. These ideas have been developed in [36], [37], [38], [39] and [40].

We shall study multiplication objects in some monoidal categories and in particular, in commutative Grothendieck categories, the category Mod-(R, τ) of the τ-injective τ-torsionfree modules and the category of graded R-modules, gr-R. The goal of this book is to deep in the study of multiplication modules and rings by means of categorical techniques. For this reason, a categorical definition is given generalizing that of multiplication module and categorical techniques are applied in order to obtain new results on multiplication modules and rings. This new definition is applied to three Grothendieck categories, namely, the category Mod-(R, τ) consisting of all τ-injective τ-torsion free R-modules, the category $\sigma[M]$ of modules subgenerated by the R-module M and the category gr-R of right graded R-modules and morphisms preserving the degree. Since it is possible to give the definition of multiplication object for an arbitrary monoidal category, there is an enormous field to apply this theory.

Since studying Grothendieck categories implies studying torsion theories over a module category, the book is organized as follows: the first chapter collects the basic tools and the necessary results to be able to establish multiplication object theory. In Section 1.1 the concept of hereditary torsion theory is defined and the main properties of closures coming from torsion theories and of modular closures are exposed. We can point out between them classical operations in Algebra such as the radical of an ideal, the S-component of an ideal (S being a multiplicative subset of the ring), the integral closure of an ideal and, of course, the closure coming from a torsion theory (this will be proved further on). The most outstanding kind of torsion theories which we have dealt with are hereditary torsion theories such as semicentered torsion theories (see e.g. [19]), torsion theories with sufficient τ-critical ideals and perfect torsion theories. They are characterized and their most important properties are exposed. Finally, the relationship between these three kinds of torsion theories are investigated, showing counterexamples in order to prove that the three kinds are different. Section 1.2 is devoted to introduce monoidal categories, emphasizing those examples which will be used on these pages.

The second chapter is devoted to the study of multiplication objects in monoidal categories. In Section 2.1 we found some examples of multiplication objects in any monoidal category and functors preserving multiplication objects are investigated. In Section 2.2, the particular case of commutative Grothendieck category is studied. Another categorical definition which does not depend on the endocommutative generator is given. The results are applied to obtain a characterization of finitely generated multiplication modules. It is proved that equivalences between commutative Grothendieck categories preserve multiplication objects and, as a result, we can give in Mod-R (R not necessarily commutative) a definition of multiplication module which will be Morita invariant. When the ring is commutative, the tensor product of a multiplication module and an invertible module is always multiplication. Multiplication objects are specially investigated in the category

$\sigma[M]$ of modules subgenerated by the R-module M in several cases. A multiplication module in $\sigma[M]$ is a multiplication object in $\sigma[M]$. The converse is solved in some cases. Multiplication objects in the category Mod-(R, τ) of τ-injective τ-torsionfree R-modules are characterized in terms of relative multiplication modules and multiplication objects in Mod-(R, τ) and Mod-R_τ (R_τ being the quotient ring with respect to τ) are related. In Section 2.3, we exemplify the theory with the cases of graded R-modules, Doi-Koppinen modules and multiplication comodules. In Section 2.4, we research into properties of endomorphisms of multiplication objects when the monoidal category is braided.

The third chapter mainly deals with the study of multiplication objects in the quotient category Mod-(R, τ). In the first section, some basic properties and concepts are established. A new definition of relative multiplication module by using semiring and semimodule theory is given. In the next sections, multiplication objects in Mod-(R, τ), or, equivalently, τ-multiplication R-modules are more deeply studied. In Section 3.2, it is verified that τ-cyclic, τ-cocritical and τ-torsion modules and τ-invertible ideals are included in this class of modules. Between the examples which appear in this chapter, it is obligatory to point out that of Krull domains as multiplication rings with respect to the canonical torsion theory, i. e., the induced one by the prime ideals with height smaller or equal to 1. For this reason, our setting seems adequate to study divisorial properties. Moreover, some kind of rings are characterized or studied in terms of multiplication rings with respect to a certain torsion theory. For example, completely integrally closed domains are characterized by using the torsion theory cogenerated by the injective hull of K/R, R being an integral domain with quotient field K. Some relative multiplication rings are characterized in terms of star-operations. Some properties related to morphisms, characterizations of τ-multiplication ideals and properties on chain conditions are exposed. Section 3.2 will be devoted to study local properties in a twofold sense. As for localization with respect to a torsion theory, it will be proved that, for perfect torsion theories, the R_τ-module M_τ (the classical module of quotients of M with respect to τ) is multiplication if and only if M is a τ-multiplication R-module. The same is not true in general. For semicentered theories, τ-finitely generated τ-multiplication R-modules are characterized as those modules M such that M_P is multiplication as an R_P-module for every τ-closed prime ideal or those that verify that M_P is cyclic as an R_P-module for prime ideals P which are τ-closed. It is proved that τ-multiplication τ-noetherian rings R are just rings which are locally (for every τ-closed prime) discrete valuation domains or SPIR's. In Section 3.3, the problem of construction new multiplication modules from others is tackled. Operations such as sum of submodules, product of τ-multiplication modules and τ-multiplication ideals, intersection and direct sum of τ-closed submodules are studied. As a consequence, all τ-multiplication modules over a Dedekind domain and over a P.I.D. (principal ideal domain) are found. It is also studied the product of τ-multiplication modules when it is possible to define it. In Section 3.4, distributive modules relative to a torsion theory are investigated, and a great number of characterizations are given. The question of when the tensor product of two τ-distributive modules or the set of morphisms between two relative distributive modules is distributive is answered. In the case of semicentered torsion theories, τ-distributive modules are characterized as those whose τ-finitely generated submodules are τ-multiplication. Other results useful to understand the relationship between relative multiplication and relative distributive module are given. Several examples show that the family of relative distributive modules is wider than that of the distributive ones.

The fourth and last chapter is concerning with the category gr-R of graded R-modules and morphisms preserving the degree. Firstly, we give basic definitions and notation and some well-known important facts of graded modules and rings. Then, examples of gr-multiplication rings which are not multiplication and all gr-multiplication domains are found. By using the Dade's equivalence between gr-R and Mod-R_e when the ring is strongly graded, it is proved that every graded R-module is gr-multiplication if and only if M_e is a multiplication R_e-module. As usual, we relate the multiplication objects in the categories gr-R, Mod-R_e and Mod-R. All these results will be obtain in Section 4.1. In Section 4.2 some local properties are exposed. It is interesting the characterization of finitely generated gr-multiplication modules as those R-modules M such that M_P^g (its localization in the graded prime P) is gr-cyclic (concept that will be defined) as an R_P^g-module for any graded prime ideal P of R. There are characterizations of gr-multiplication modules in terms of their behavior with respect to product and intersections of a family of graded ideals. Properties such as being cofinitely generated in gr-R are studied for multiplication objects. In Section 4.3, gr-distributive modules are studied and it is proved that they are just graded R-modules M such that their finitely generated submodules are gr-multiplication, or, equivalently, for every graded prime ideal P, M_P^g is a gr-cyclic R_P^g-module. Section 4.4 is concerning with multiplication graded rings. Several local characterizations are obtained for gr-multiplication rings. Multiplication graded domains are exactly gr-Dedekind domains. The graded notions of von Neumann regular ring (gr-regular), SPIR or discrete valuation domain will be frequently used. When the grading group is ordered, a graded version of Mott's Theorem is found. In Section 4.5, we also answer the question of when a graded ring is multiplication. For example, it is proved that integral domains strongly graded by $\mathbb{Z}$ are exactly group rings of the form $K[\mathbb{Z}]$, K being a field. Noetherian multiplication rings graded by a torsion free group are determined. If moreover, the grading group is $\mathbb{Z}$, then R is multiplication if and only if R_e is a finite product of fields. It is proved that, for noetherian rings positively graded by $\mathbb{Z}$, being gr-multiplication and being multiplication are equivalent properties. It is also investigated the case of commutative twisted group algebras when the grading group is finite and the problem is totally solved when the field is perfect.

In order to finish this introduction, I would like to thank Professor Blas Torrecillas Jover, from Almería University, for giving me the chance to research into these subjects and for his dedication, understanding and encourage.

I thank Professor Patrick F. Smith, from Glasgow University, his ideas, suggestions and continuous support, in particular, for my stay in Glasgow University.

I wish to show my gratitude to Professor Constantin Năstăsescu, from Bucarest University, for his valuable comments, expressed in several conversations maintained for his stay in Almería University.

Last, I owe the greatest debt to David, Daniel and Begoña.

Chapter 1

Torsion theories and monoidal categories

Every Grothendieck category is equivalent, by Gabriel-Popescu theorem, to a quotient category of a module category with respect to a hereditary torsion theory. For this reason, before explaining the elementary concepts on monoidal and commutative Grothendieck categories, we shall expose properties of hereditary torsion theories and, in particular, of semicentered and perfect torsion theories. We shall also deal with modular closure properties (Kirby closure), as they are related to closures coming from a torsion theory. We consider always (unless stated otherwise) that rings are commutative with an identity element and modules are unital.

We make explicit concepts and basic properties on hereditary, semicentered and perfect torsion theories. It will be proved that every closure coming from a torsion theory is a componental operation and therefore, a modular closure. For it, this particular kind of closure operation is investigated. This result is proved by using torsion theories techniques and not completing or generalizing the work by Chew (cf. [26]). In the section devoted to monoidal categories, after that giving the definition, several examples are showed in order to prove that it is an important kind of category. It is pointed out that every commutative Grothendieck category has associated in a natural way a monoidal category that is (monoidal) equivalent to a category of the kind Mod-(R, τ).

1.1 Commutative torsion theories

By associating to Mod-R a set of torsion theories and by using their properties, in many cases, it is possible to determine properties of Mod-R and R. These techniques can be observed in [7], [53] or [105].

Recall the definition of torsion theory which appears in [105, Chapter 6] and it is due to Dickson.

Definition 1.1.1 *A torsion theory τ on R-modules is a pair $\tau = (\mathbf{T}, \mathbf{F})$ of classes of R-modules such that*

a) $Hom(T, F) = 0$ for every $T \in \mathbf{T}$ and $F \in \mathbf{F}$.

b) If $Hom(C, F) = 0$ for every $F \in \mathbf{F}$, then $C \in \mathbf{T}$.

c) If $Hom(T, C) = 0$ for every $T \in \mathbf{T}$, then $C \in \mathbf{F}$.

Modules belonging to **T** and **F** are called *τ-torsion* and *τ-torsion free* respectively. As it can be seen in [105, Propositions 2.1 and 2.2], **T** is closed under quotient objects, coproducts and extensions (given any short exact sequence of R-modules with ends being τ-torsion, the central term is τ-torsion as well), and $\mathbf{F}$ is closed under subobjects, products and extensions. If, moreover, $\mathbf{T}$ is closed under subobjects, then τ is called hereditary.

For every R-module M, there exists a largest τ-torsion submodule which will be denoted by $\tau(M)$. The assignment $M \mapsto \tau(M)$ is functorial.

Recall that if A and B are ideals of R, then $(A : B) = \{r \in R; Br \subseteq A\}$. In the particular case of the ideal aR $(a \in R)$, we can write $(A : a)$ instead of $(A : aR)$.

Definition 1.1.2 *A Gabriel topology on a commutative unitary ring R is a topology defined by a family $\mathcal{F}$ of ideals such that:*

G1.- $A \in \mathcal{F}$, $A \subseteq B \Rightarrow B \in \mathcal{F}$.

G2.- $A, B \in \mathcal{F} \Rightarrow A \cap B \in \mathcal{F}$.

G3.- $A \in \mathcal{F} \Rightarrow$ $(A : a) \in \mathcal{F}$ for all $a \in R$.

G4.- If A is an ideal of R and there exists $B \in \mathcal{F}$ such that $(A : b) \in \mathcal{F}$ for all $b \in B$ then $A \in \mathcal{F}$.

Hereditary torsion theories are in bijective correspondence with Gabriel topologies ([105, 6.5.1]). In [105, 6.5.2 and 6.5.3] it is shown that G3 and G4 imply G1 and G2 and that $\mathcal{F}$ is closed under products. $\mathcal{F}$ is also called a *Gabriel filter.*

In the commutative case, a Gabriel filter can be defined by $G4$ and the condition $R \in \mathcal{F}$.

Let M be an R-module and $\mathcal{F}$ a Gabriel filter in the ring R, with τ =($\mathbf{T}$,$\mathbf{F}$) the corresponding hereditary torsion theory. The τ*-closure* of $N \leq M$ is $Cl_\tau^M(N) = \{x \in M;\ \exists I \in \mathcal{F}) \mid x.I \subseteq N\}$.

It is obvious that $Cl_\tau^M(N) = \{x \in M; (N : x) \in \mathcal{F}\}$ where $(N : x) = \{r \in R; xr \in N\}$. A submodule N of M is said to be *a τ-closed submodule* if it coincides with its closure. According to [90, Lemma 2.1], if τ is a hereditary torsion theory on a commutative unitary ring R and P is a prime ideal of R, then P is τ-closed if and only if P does not belong to the Gabriel filter. So, there is a partition in the set of all prime ideals. The set of all τ-closed prime ideals is denoted by $Spec_\tau(R) = \{P \in Spec(R); \tau(R/P) = 0\}$ or $K(\tau)$ (cf. [19]).

The next proposition collects well-known properties of the closure operation defined by a torsion theory (cf. [26], [53] or [105]). Recall that if $\mathcal{L}$ is a lattice, a *lattice closure operation* is a map $c : \mathcal{L} \to \mathcal{L}$ verifying the three following properties:

1. $A \subseteq c(A) \ \ \forall A \in \mathcal{L}$.
2. $c(c(A)) = c(A) \ \ \forall A \in \mathcal{L}$.
3. $A \subseteq B \Rightarrow c(A) \subseteq c(B) \ \ \forall A, B \in \mathcal{L}$.

Proposition 1.1.3 *Let R be a unitary commutative ring. Let τ be a torsion theory on R. Let M be an R-module. Then*

C1.- Cl_τ^M is a closure operation of lattices.

C2.- Let L and N be submodules of M such that $L \subseteq N$. Then $Cl_\tau^M(L) \cap N = Cl_\tau^N(L)$

C3.- $Cl_\tau^M(L) \cap Cl_\tau^R(N) = Cl_\tau^M(L \cap N)$ for all $L, N \leq M$.

C4.- $Cl_\tau^R(I) \subseteq \cap_{P \in \mathcal{A}} P$ for all $I \leq R$ where $\mathcal{A} = \{P \in Spec_\tau(R); I \subseteq P\}$.

C5.- $Cl_\tau^M(Cl_\tau^M(L) + Cl_\tau^M(N)) = Cl_\tau^M(L + N)$ for all $L, \ N \leq M$

C6.- $Cl_\tau^R(I) = I \Rightarrow Cl_\tau^R(I : J) = (I : J)$ *for all* $I, J \leq R$ *(the division is made in* R*).*

The following properties are less known and therefore the proof is given. The situation is the same as the above proposition. Recall that a ring R is τ-*noetherian* if the lattice of τ-closed ideals of R is noetherian. Equivalent conditions appear in [105, Proposition XIII.2.4]. An ideal I of R is τ-finitely generated if $I = Cl_\tau^I(J)$ for some finitely generated ideal $J \subseteq I$. In the same way τ-finitely generated submodule and τ-noetherian module are defined. In the next proposition, when we write $N_P \cap M$ we mean $\{x \in M; \frac{x}{1} \in N_P\}$.

Proposition 1.1.4 *Let* τ *be a hereditary torsion theory on* R *and* M *an* R*-module. The following properties are satisfied.*

C7.- Let $L,\ N \leq M$*. Let* $J \in \mathcal{F}$*. If* $Cl_\tau^M(L.J) \subseteq Cl_\tau^M(N.J)$ *then* $Cl_\tau^M(L) \subseteq Cl_\tau^M(N)$.

C8.- $Cl_\tau^R(Cl_\tau^R(I).Cl_\tau^R(J)) = Cl_\tau^R(I.J)$ *for all* $I, J \leq R$.

C9.- If R *is* τ*-noetherian then* $Cl_\tau^M(N) = \cap_{P \in Spec_\tau(R)}(N_P \cap M)$*, for all* $N \leq M$.

Proof:

C7.- Consider $x \in Cl_\tau^M(L)$. Then there exists $H \in \mathcal{F}$ such that $xH \subseteq L$. Hence $xHJ \subseteq LJ$. Thus $Cl_\tau^M(NJ) \supseteq Cl_\tau^M(LJ) \supseteq xJ$. Hence there exists $H' \in \mathcal{F}$ such that $xH'J \subseteq NJ \subseteq N$. Therefore, $x \in Cl_\tau^M(N)$.

C8.- We have

$$IJ \subseteq Cl_\tau^R(I).Cl_\tau^R(J) \Rightarrow Cl_\tau^R(IJ) \subseteq Cl_\tau^R(Cl_\tau^R(I)Cl_\tau^R(J)).$$

Moreover, if $x \in Cl_\tau^R(Cl_\tau^R(I)Cl_\tau^R(J))$ then there exists $L \in \mathcal{F}$ such that $xL \subseteq Cl_\tau^R(I)Cl_\tau^R(J)$. We will see that $Cl_\tau^R(I)Cl_\tau^R(J) \subseteq Cl_\tau^R(IJ)$. In fact,

$$y \in Cl_\tau^R(I)Cl_\tau^R(J) \Rightarrow y = a_1b_1 + ... + a_nb_n,\ a_i \in Cl_\tau^R(I),\ b_i \in Cl_\tau^R(J).$$

So, $\exists\ L_i, L'_j$ such that $a_iL_i \subseteq I, b_jL'_j \subseteq J$ for all $i, j \in \{1, ..., n\}$.
We consider $L = L_1 \cap ... \cap L_n$ and $L' = L'_1 \cap ... \cap L'_n$. Hence $LL' \in \mathcal{F}$ and $yLL' \subseteq IJ$. So, $y \in Cl_\tau^R(IJ)$. Hence, $x \in Cl_\tau^R(Cl_\tau^R(IJ)) = Cl_\tau^R(IJ)$.

C9.-We take $x \in Cl_\tau^M(N)$. Then there exists $H \in \mathcal{F}$ such that $xH \subseteq N$. Then $(xH)_P = xR_P$ for all $P \in Spec_\tau(R)$ because $(xH)R_P = x(HR_P) = xR_P \subseteq NR_P$ as H is not contained in P (otherwise P would belong to the filter.)

Let $x \in \cap_{P \in Spec_\tau(R)}(NR_P)$. Then $x/1 \in NR_P$ for all $P \in Spec_\tau(R)$.

This implies that there exists $s_p \in R - P$ such that $xs_p \in N$ for all $P \in Spec_\tau(R)$.
So, $(N : x)$ is not contained in P and then $(N : x)_P = R_P$ for all $P \in Spec_\tau(R)$. Then $Cl_\tau^R(N : x) = R$ and therefore $(N : x) \in \mathcal{F}$. It is clear that $(N : x) \subseteq N$. Hence $x \in Cl_\tau^M(N)$.■

Modular closures are studied to characterize axiomatically classical maps in Algebra as the integral closure of an ideal, the S-component of an ideal (where S is a multiplicatively closed subset of the ring) and the radical of an ideal (if I is an ideal of R, its *radical* is $\{r \in R; r^n \in I$ for some $n \in \mathbb{Z}^+\})$. We shall express basic properties and characterizations, which will be frequently used further on.

The following results are taken from [67].

Definition 1.1.5 *Let M be an R-module, and $L(M)$ the lattice of submodules. A Kirby closure operation (or modular closure) is a mapping $g : L(M) \longrightarrow L(M)$ which satisfies the following properties:*

K1.- $L \subseteq g(L)$ for all $L \in L(M)$.

K2.- $g(g(L)) = g(L)$ for all $L \in L(M)$.

K3.- $L \subseteq N \Rightarrow g(L) \subseteq g(N)$ for all $L, N \in L(M)$.

K4.- $g(L).A \subseteq g(L.A)$ for all $A \leq R,\ L \in L(M)$.

Proposition 1.1.6 *A mapping $g : L(M) \longrightarrow L(M)$ is a Kirby closure if and only if $(g(L) : g(N)) = (g(L) : N)$ for all $L,\ N \in L(M)$.*

Properties of Kirby closures.

PK1.- $g(\cap_{\lambda\in\Lambda} L_\lambda) \subseteq \cap_{\lambda\in\Lambda} g(L_\lambda) = g(\cap_{\lambda\in\Lambda} g(L_\lambda))$ for all $L_\lambda \in L(M)$.

PK2.- $\sum_{\lambda\in\Lambda} g(L_\lambda) \subseteq g(\sum_{\lambda\in\Lambda} L_\lambda) = g(\sum_{\lambda\in\Lambda} g(L_\lambda))$ for all $L_\lambda \in L(M)$.

PK3.- $g(L : A) \subseteq (g(L) : A) = g(g(L) : A)$ for all $L \in L(M)$ and $A \leq R$.

PK4.- $g(L.A) = g(g(L).A)$ for all $A \leq R,\ L \in L(M)$.

The componental operation is a particular case of Kirby closure. The following results are taken from [67].

Definition 1.1.7 *A componental operation is a mapping $g : L(M) \longrightarrow L(M)$ such that it is a Kirby closure and verifies:*

$g(m_2.(L : m_1.R)) \supseteq m_2.(g(L) : m_1.R)$ for all $L \in L(M)$ and for all $m_1, m_2 \in L(M)$.

Corollary 1.1.8 *If $\mathcal{I}$ is the ideal lattice of a unitary commutative ring R, then $f : \mathcal{I} \longrightarrow \mathcal{I}$ is a componental operation if and only if f is a Kirby closure and $f(A : aR) = (f(A) : aR)$ for all $A \in \mathcal{I}$, $a \in R$.*

Properties of componental operations.

PC1.- $g(L_1) \cap ... \cap g(L_n) = g(L_1 \cap ... \cap L_n)$ for all $L_1, ..., L_n \in L(M)$.

PC2.- $g(L : A) = (g(L) : A)$, if A is finitely generated, for all $L \in L(M)$ and $A \in R$.

PC3.- $g(C(L : N)) = g(C)(g(L) : N)$, if $N \in L(M)$ is finitely generated, for every $L, C \in L(M)$.

Lemma 1.1.9 *Each closure coming from a hereditary torsion theory is a componental operation and therefore, a modular closure.*

Proof: We will prove the four properties of a Kirby closure.

a) $N \subseteq Cl_\tau^M(N)$ for all $N \leq M$.

Clear.

b) $N \leq N' \Rightarrow Cl_\tau^M(N) \subseteq Cl_\tau^M(N')$ for all $N, N' \leq M$.

Clear.

c) $Cl_\tau^M(Cl_\tau^M(N)) = Cl_\tau^M(N)$ for all $N \leq M$.

By a) and b) the inclusion $Cl_\tau^M(N) \subseteq Cl_\tau^M(Cl_\tau^M(N))$ is trivial.

Take $x \in Cl_\tau^M(Cl_\tau^M(N))$. Then there exists $I \in \mathcal{L}(\tau)$ such that $x.I \subseteq Cl_\tau^M(N)$.

Consider $i \in I$. As $x.i \in Cl_\tau^M(N)$, then there exists $J_i \in \mathcal{F}$ such that $x.i.J_i \subseteq N$. Therefore $((N : x) : i) = (N : x.i) \supseteq J_i \in \mathcal{F}$ and hence $((N : x) : i) \in \mathcal{F}$ and by G4 $(N : x) \in \mathcal{F}$. So, $x(N : x) \subseteq N$ and $x \in Cl_\tau^M(N)$.

d) $Cl_\tau^M(N).A \subseteq Cl_\tau^M(N.A)$ for all $A \leq R,\ N \leq M$.

We take $x \in Cl_\tau^M(N).A$. Then $x = y_1.a_1 + ... + y_n.a_n$ with $a_i \in A$ and $y_i \in Cl_\tau^M(N)$. So, $\exists I_i \in \mathcal{F}$ such that $y_i.I_i \subseteq N$. Therefore, $I = I_1 \cap ... \cap I_n \in \mathcal{F}$ and $x.I \subseteq N.A$. This implies $x \in Cl_\tau^M(N.A)$.

Let $m_1, m_2 \in M$. Let $N \leq M$. If $x \in m_2.(Cl_\tau^M(N) : m_1.R)$, then $x = m_2.y$ where $m_1.y \subseteq Cl_\tau^M(N)$. Therefore there exists $I \in \mathcal{F}$ such that $m_1.y.I \subseteq N$. Hence $y.I \subseteq (N : m_1.R)$. Thus $x.I = m_2.y.I \subseteq m_2.(N : m_1.R)$. Consequently, $x \in Cl_\tau^M(m_2.(N : m_1.R))$ and the definition of componental operation is verified. Thus the closure is a componental operation.■

This result was proved by Chew (cf. [26]) in a different way.

Inside hereditary torsion theories, the semicentered ones (they are also called *primal*) will be the most useful in this work. Occasionally, perfect torsion theories, which are a particular case of the aforementioned theories, will be taken into account. We will define and characterize semicentered torsion theories. Rings where every torsion theory is of this kind are studied in Pakala's thesis (cf. [96]) and they are called *semicentered rings.* In several Lemmas, some technical properties are collected, which will be used very often.

Let τ be a hereditary torsion theory on R and let M be any R-module.

Definition 1.1.10 *An R-module M is called τ-cocritical if is τ-torsion free and every quotient module different from M is τ-torsion.*

Definition 1.1.11 *An ideal I of R is called τ-critical if R/I is τ-cocritical.*

It is clear that if R is a commutative ring, then $P \in Spec(R)$ is τ-critical if and only if P is maximal with respect to the condition of not belonging to the Gabriel filter, $\mathcal{F}$. The set of all prime τ-critical ideals will be denoted by $Max_{\mathcal{F}}(R)$ or $Max_{\tau}(R)$.

Remark. Any element of $Max_{\mathcal{F}}(R)$ is a prime ideal (cf. [105, Lemma VI.6.14]).

Definition 1.1.12 *If for every ideal $A \notin \mathcal{F}$ there exists a τ-critical ideal, P, such that $A \subseteq P$, then we say that τ has sufficient τ-critical ideals over R or that R has sufficient τ-critical ideals.*

Torsion theories of finite kind are examples of torsion theories having sufficient τ-critical ideals over R (cf. [15, Theorem 3.1]).

The following definition comes from [19].

If $K \subseteq Spec(R)$, then the set, $\mathcal{F}(K)$, of ideals I of R such that $K \subseteq \{P \in Spec(R); I \nsubseteq P\}$ is a Gabriel filter. Conversely, if τ is a hereditary torsion theory, then we can consider $K(\tau) = Spec(R) - \mathcal{F}$.

Definition 1.1.13 *A torsion theory τ on R is called semicentered if $\mathcal{F}(K(\tau)) = \mathcal{F}$.*

Definition 1.1.14 *A ring R is called semicentered if every torsion theory on R is semicentered.*

Noetherian rings, semiartinian rings and discrete valuation domains are examples of semicentered rings (cf. [96]).

If τ has sufficient τ-critical ideals on R, then τ is semicentered but the converse is false (cf. [15, Theorem 3.1]).

Lemma 1.1.15 *A hereditary torsion theory τ is semicentered if and only if for every $I \notin \mathcal{F}$ there exists $P \in K(\tau)$ such that $I \subseteq P$.*

Proof: It is [19, Proposition 3.3.15]. ■

A torsion theory is of *finite type* if the Gabriel filter has a basis consisting of finitely generated ideals. Every torsion theory of finite type is semicentered according to [19, Proposition 3.3.17]. The next Lemma collects some useful technical results. We clarify some concepts before enunciating the Lemma. If τ is a torsion theory over R and S is a multiplicatively closed subset of R, then $S^{-1}\tau$ is a torsion theory over the ring of quotients $S^{-1}R$ whose Gabriel filter consists of the localized ideals of ideals belonging to the the filter of τ. When $S = R-P$, P being a prime ideal, $S^{-1}\tau$ is denoted by τ_P. A submodule L of M is called *S-saturated* if it verifies that if $x \in M$ and $xs \in L$ for some $s \in S$, then $x \in L$. There is a bijection between S-saturated submodules of M and $S^{-1}R$-submodules of $S^{-1}M$, given by the maps which map L to $S^{-1}L$ and N to $f^{-1}(N)$, where $f : M \to S^{-1}(M)$ is the canonical homomorphism (further details can be found in [105]).

Lemma 1.1.16 *Let S be a multiplicatively closed subset of R. Let τ be a semicentered in Mod-R. Let $P \in K(\tau)$. Let M, N be two R-modules and let $L \leq M$. Then,*

1. $S^{-1}Cl_\tau^M(L) = Cl_{S^{-1}\tau}^{S^{-1}M}(S^{-1}(L))$.

2. $Cl_{\tau_P}^{M_P}(L_P) = L_P$.

3. *if N is τ-finitely generated, then $(M : N)_P = (M_P : N_P)$ for every $P \in K(\tau)$.*

4. *if M is τ-finitely generated, then $(ann(M))_P = ann(M_P)$ for every $P \in K(\tau)$.*

Proof: Every element of $S^{-1}Cl_\tau^M(L)$ is of the form $\frac{x}{s}$ with $s \in S$ and $x \in Cl_\tau^M(L)$. Thus there exists an ideal $I \in \mathcal{F}$ such that $xI \subseteq L$. Hence, $\frac{x}{s}I \subseteq L$ and by applying S^{-1}, it is obtained $\frac{x}{s}S^{-1}(I) \subseteq S^{-1}(L)$. Therefore, $S^{-1}Cl_\tau^M(L) \subseteq Cl_{S^{-1}\tau}^{S^{-1}(M)}(S^{-1}(L))$. In order to prove the other inclusion, take an element $\frac{m}{s} \in Cl_{S^{-1}\tau}^{S^{-1}(M)}(S^{-1}(L))$. Then, there exists an ideal $I \in \mathcal{F}$ such that $\frac{m}{s}(S^{-1}I) \subseteq S^{-1}(L)$. Hence, for every $i \in I$ there exist $l \in L$ and $t \in S$ such that $\frac{mi}{s} = \frac{l}{t}$. Therefore, there exists $z \in S$ such that $mzi \in L$. If we take L S-saturated, it follows that $mi \in L$. It is deduced that $m \in Cl_\tau^M(L)$ and the proof of the first point ends.

Now, we prove the second one. Since the inclusion is a closure operation, $L_P \subseteq Cl_{\tau_P}^{M_P}(L_P)$ holds always. In this case $S = R - P$. Let $\frac{x}{s} \in Cl_{\tau_P}^{M_P}(L_P)$. Then, there exists $I \in \mathcal{F}$ such that $\frac{x}{s}(S^{-1}(I)) \subseteq L_P$. Since $P \in K(\tau)$, there exists $i \in I \cap (R - P)$ such that $\frac{xi}{s} = \frac{l}{t}$ for some $l \in L$ and $t \in S$. Hence, there exists $z \in S$ such that $xz \in L$ and by taking L S-saturated, it follows that $x \in L$ and therefore, $\frac{x}{s} \in L_P$.

For the next point, by hypothesis, $N = Cl_\tau^N(F)$, with F finitely generated. Then, it holds $(M : N)_P \subseteq (M : F)_P = (M_P : F_P) = (M_P : N_P)$, where it has been used that F is finitely generated and the point 2 of this Lemma. Let $\frac{r}{s} \in (M_P : N_P)$ and let $n \in N$. Since $\frac{nr}{s} \in M_P$, it follows that there exists $z \in S$ such that $nzr \in M$. Thus, given $n \in N$, $\frac{r}{s}$ can be written as $\frac{rz}{sz}$, where $rz \in (M : N)$ and therefore, $\frac{r}{s} \in (M : N)_P$.

To prove the last point, let $F =< f_1, ..., f_n >$ finitely generated such that $M = Cl_\tau^M(F)$. Let $\frac{r}{s} \in [ann(M)]_P$ and $m \in M$. Since $r \in ann(M)$, $\frac{mr}{s} = 0$,

i. e., $[ann(M)]_P \subseteq ann(M_P)$ is an inclusion that is always true.

Let $\frac{r}{s} \in [ann\ (M)]_P$ and let $m \in M$. Since $M = Cl_\tau^M(F)$, there exists $I \in \mathcal{F}$ such that $mI \subseteq F$ and since $P \in \mathcal{F}$, there exists $i \in I \cap (R-P)$ with $m.i \in F$. On the order hand, since $r \in ann(M_P)$, $\frac{f_i r}{s} = 0$ $(1 \leq i \leq n)$. It is deduced that exists $z \in S = (R-P)$ such that $frz = 0$ for every element $f \in F$. Then, $\frac{r}{s}$ can be expressed as $\frac{riz}{siz}$ and, therefore, $mirz = 0$ as $mi \in F$. In this way, the other inclusion holds. ■

Denote by R_τ the *ring of quotients of R with respect to the torsion theory* τ in the sense of [105, Chapter 6]. Then, R_τ is the direct limit taken over the ideals I belonging to the Gabriel filter with the order converse to the inclusion of the sets $Hom_R(I, R/\tau(R))$ where $\tau(R)$ is the torsion submodule of R. Analogously, it is defined the *module of quotients* of M respect to τ, M_τ.

Definition 1.1.17 *A hereditary torsion theory τ over R is called perfect if the canonical homomorphism i from R to R_τ is a ring epimorphism making R_τ into a flat R-module and the Gabriel filter is $\mathcal{F} = \{I \leq R; R_\tau i(I) = R_\tau\}$.*

Remark. If τ is perfect, by [105, 11.3.4], the filter has a basis of finitely generated ideals and therefore, τ is of finite type. Hence, τ has sufficient τ-critical ideals and by [15, Theorem 3.1], it is semicentered.

Remark. It is convenient to emphasize that [105, Chapter XI] gives the following:

1. If the Gabriel filter has a basis consisting of τ-invertible ideals (an R-submodule I of R_τ is called *τ-invertible* if there exist $a_1, ..., a_n \in I$ and

$q_1, ..., q_n \in I^* = \{q \in R_\tau; Iq \subseteq R\}$ such that $1 = \sum_{i=1}^{i=n} a_i q_i)$, then τ is perfect (11.3.5).

2. If R is a hereditary noetherian ring, then every hereditary torsion theory is perfect (11.3.6).

3. If R is a commutative ring, then a 1-topology (Gabriel topology with a basis of principal ideals) is always perfect (11.6.3).

4. If R is a commutative ring and S is a multiplicatively closed subset of R, then $S^{-1}R = R_\tau$, where the Gabriel filter is $\{I \leq R; I \cap S \neq \emptyset\}$, and τ is perfect (11.6.4).

The following examples show that the next implications are strict.
τ is perfect $\Rightarrow$ τ has sufficient τ-critical ideals on $R \Rightarrow \tau$ is semicentered

Example 1.1.18 Consider a valuation domain R with order of value group is order isomorphic to the additive group of rational numbers with the classical order. If M is the maximal ideal of R, then $M^2 = M$ and $Spec(R) = \{0, M\}$. Thus $\{M, R\}$ is a Gabriel filter is not of the form $\mathcal{F}_\mathcal{P} = \{I \leq R; I_P = R_P \ \forall P \in \mathcal{P}\}$ for every $\mathcal{P} \subseteq Spec(R)$. By [105, 6.6.13], if τ is semicentered, then its Gabriel filter has the form $\mathcal{F}_\mathcal{P}$ for some $\mathcal{P} \subseteq Spec(R)$. Therefore, there are hereditary torsion theories that are not semicentered.

This example appears in [18].

Example 1.1.19 Let K be a field and S the polynomial ring on K in the indeterminates $X_1, X_2, ...$ that commute between them. Let N be the ideal generated by all $X_i's$ and let R be the localization of S in the complement of N and let M be the only maximal ideal of R. For every $n = 1, 2, ...,$ let P_n be the prime ideal of R generated by $X_1, X_2, ..., X_n$. Let τ_i be the torsion theory such that its Gabriel filter $\mathcal{F}_i$ consists of all ideals which are not contained P_i and let $\mathcal{F}$ the intersection of $\mathcal{F}_i's$. I is in $\mathcal{F}$ if and only if for any n, I is not contained in P_n.

Suppose that there is an ideal I maximal with respect to the exclusion of $\mathcal{F}$. Then, I is contained in some P_n, and since P_n is not contained in $\mathcal{F}$ and I has the indicated maximality property, $I = P_n$. But P_{n+1} contain properly P_n and it is not in $\mathcal{F}$, which contradicts the maximality of I. There are no ideals I such as the described ones. $\mathcal{F}$ does not have sufficient τ-critical ideals but is semicentered.

This example and the next one appear in [15].

Example 1.1.20 Let K be a field, S be a polynomial ring on K in the indeterminates Y_t, $t \in [0\ ,\ 1]$ which commute. Let A be the ideal of S generated by $Y_t^2 - Y_t$, $t \in [0\ ,\ 1]$. Let $R = S/A$. By X_t we denote the image of Y_t in R, M is the maximal ideal of R generated by the $X_t's$. Then, $M^2 = M$ and $\mathcal{F} = \{R, M\}$.

Suppose that I is an ideal of R but it does not belong to $\mathcal{F}$. Then, X_s is not in I for some $s \in [0\ ,\ 1]$. Let $J = I + (1 - X_s)R$. Since $1 - X_s$ is in J, J is not contained in M. It holds $J \neq R$ as otherwise $a + (1 - X_s)r = 1$ with $a \in I$; by multiplying by X_s we obtain that $aX_s = X_s$, and hence, X_s is in I, a contradiction. Therefore, J is a proper ideal that is not contained in M, and it is contained in a maximal ideal of R not belonging to $\mathcal{F}$. As a consequence, R has sufficient τ-critical ideals. However, this torsion theory is not of finite type and for this reason, it is not perfect (cf. [15, Example 3.3]).

Lemma 1.1.21 *If τ is perfect on R, then $P \in Max_{\mathcal{F}}(R)$ if and only if $i(P)R_\tau$ is a maximal ideal of R_τ. Moreover, every maximal ideal of R_τ has this form.*

Proof: Consider $P \in Max_{\mathcal{F}}(R)$. Then, $P \notin \mathcal{F}$. Hence, $i(P)R_\tau$ is different from R_τ. By [105, 11.3.8], there is a bijective correspondence between ideals of R_τ and the set $S = \{I \leq R; I = i^{-1}(i(I)R_\tau)\}$ given by $b(V) = i^{-1}(V)$. Then, $i^{-1}(i(P)R_\tau \notin \mathcal{F}$ because otherwise, $i(i^{-1}(i(P)R_\tau))R_\tau = R_\tau = (i(P)R_\tau)R_\tau = i(P)R_\tau$, a contradiction.

Moreover, $i(P) \subseteq i(P)R_\tau$ and $P \subseteq i^{-1}(i(P)R_\tau)$. Since $i^{-1}(i(P)R_\tau) \notin \mathcal{F}$, by the maximality of P, $P = i^{-1}(i(P)R_\tau)$. If $i(P)R_\tau \leq V$, then, by applying i^{-1}, we have $P \leq i^{-1}(V)$ and this ideal is different from R because V is different from R_τ.

Now, consider a maximal ideal V of R_τ. Then, $i^{-1}(V) \notin \mathcal{F}$. Otherwise, $i(i^{-1}(R))R_\tau = VR_\tau = V = R_\tau$, a contradiction.

If we suppose that there exists $P \in Max_{\mathcal{F}}(R)$ such that $i^{-1}(V) \subseteq P$, then by applying b^{-1}, we obtain $V.R_\tau = i(P)R_\tau$, which is different from R_τ. Hence, $i^{-1}(V) = P$ y $V = i(P)R_\tau$.■

1.2 Monoidal categories

A category $\mathcal{C}$ is said to be a *monoidal category* (other authors call it *tensor category*) if there is a bifunctor $\otimes_{\mathcal{C}} : \mathcal{C} \times \mathcal{C} \to \mathcal{C}$ (called *tensor product*), an object I (called *the unit* of the monoidal category) and natural isomorphisms $a_{\{A,B,C\}} : (A \otimes_{\mathcal{C}} B) \otimes_{\mathcal{C}} C \to A \otimes_{\mathcal{C}} (B \otimes_{\mathcal{C}} C)$, $l_A : I \otimes_{\mathcal{C}} A \to A$ and $r_A : A \otimes_{\mathcal{C}} I \to A$ for each A, B and C in $\mathcal{C}$, verifying the following four formulas, which will be briefly denoted by (1), (2), (3) and (4) and where $f : A \to A'$, $g : B \to B'$ and $h : C \to C'$ represent morphisms in $\mathcal{C}$.

$$f \otimes_{\mathcal{C}} (g \otimes_{\mathcal{C}} h) \circ a_{\{A,B,C\}} = a_{\{A',B',C'\}} \circ (f \otimes_{\mathcal{C}} g) \otimes_{\mathcal{C}} h$$

$$l_{A'} \circ (1_I \otimes_{\mathcal{C}} f) = f \circ l_A, \quad r_{A'} \circ (f \otimes_{\mathcal{C}} 1_I) = f \circ r_A$$

$$(1_A \otimes_{\mathcal{C}} l_B) \circ a_{\{A,I,B\}} = r_A \otimes_{\mathcal{C}} 1_B$$

$$a_{\{A,B,C\otimes_{\mathcal{C}}D\}} \circ a_{\{A\otimes_{\mathcal{C}}B,C,D\}} = (1_A \otimes_{\mathcal{C}} a_{\{B,C,D\}}) \circ a_{\{A,B\otimes_{\mathcal{C}}C,D\}} \circ (a_{\{A,B,C\}} \otimes_{\mathcal{C}} 1_D)$$

Property (3) is known as *triangle axiom* and property (4) is called *pentagon axiom*. In case of the isomorphisms $a_{\{A,B,C\}}, l_A$ and r_A are identities, $\mathcal{C}$ is said to be a *strict monoidal category*. The above defined category will be denoted by $(\mathcal{C}, \otimes_{\mathcal{C}}, I, a, l, r)$.

It is well-known that, in every monoidal category,

$$l_I = r_I, \quad (1_A \otimes_{\mathcal{C}} r_B) \circ a_{\{A,B,I\}} = r_{A\otimes_{\mathcal{C}}B}, \quad l_A \otimes_{\mathcal{C}} 1_B = l_{A\otimes_{\mathcal{C}}B} \circ a_{\{I,A,B\}}$$

(see, for example, [64, Proposition 1.1]). Another important fact for our purposes appears in [66, Proposition XI.2.4], where it is stated that the set of endomorphisms of the unit, $End(I)$, is a commutative monoid with the composition. Moreover, identifying $I \otimes_{\mathcal{C}} I$ to I, the tensor product coincides, in $End(I)$, with the composition.

Given two monoidal categories, $\mathcal{C}$ and $\mathcal{D}$, with units I and I' respectively, a functor between them, F, is said to be a *monoidal functor* if there exists an isomorphism $\Phi_0 : I' \to FI$ and $\Phi_{2,A,B} : FA \otimes FB \to F(A \otimes B)$, a family of isomorphisms verifying the following equalities:

fu1. $\Phi_{2,A,B\otimes C} \circ (1_{FA} \otimes \Phi_{2,B,C}) \circ a_{\{FA,FB,FC\}} = F(a_{\{A,B,C\}}) \circ \Phi_{2,A\otimes B,C} \circ (\Phi_{2,A,B} \otimes 1_{FC})$.

fu2. $r_{FA} = F(r_A) \circ \Phi_{2,A,I} \circ (1_{FA} \otimes \Phi_0)$.

fu3. $l_{FA} = F(l_A) \circ \Phi_{2,I,A} \circ (\Phi_0 \otimes 1_{FA})$.

When the functors intervening in an equivalence of categories are monoidal functors, the categories are called *monoidal equivalent.* Notice that the two tensor products have been denoted in the same way.

Examples. We expose some examples of monoidal categories, most of them are well-known.

The category $Vect(K)$ of vector spaces over a field K, with $I = K$ and the tensor product being that of vector spaces. The mappings l and r are defined by $l(1 \otimes v) = r(v \otimes 1) = v$ for all $v \in V$ (V being an object in the category). The category Mod-$K[G]$ of representations of G over K or, equivalently, of $K[G]$-modules is a submonoidal category of $Vect(K)$ with the structures $g(u \otimes v) = gu \otimes gv$ and $gk = k$ for all $g \in G$, $u, v \in V$ and $k \in K$.

Any monoid, regarded as a discrete category is a strict monoidal category, where the tensor product is the multiplication.

The category *Ab* of abelian groups with the usual tensor product and with $\mathbb{Z}$ as the unit.

The category of right R-modules, Mod-R, when R is commutative, with the usual tensor product and R being the unit.

The category of K-algebras with the tensor product of K-algebras and K being the unit.

The category of all R-R-bimodules, R being any ring with the tensor product in R.

The opposite category to any monoidal category.

The category $({}_B\mathcal{M}, \otimes, K, a, l, r)$ of left B-modules, B being a bialgebra, is monoidal.

Let gr-R be the category consisting of all R-modules graded by a group G, R being a commutative ring and morphisms preserving the degree. Then, the category gr-R is a monoidal category considering R as the unit, defining the graduation for the tensor product of two R-modules, M and N, by means of $(M \otimes N)_g$, which is the additive subgroup generated by the elements $x \otimes y$ with $x \in M_h$, $y \in N_l$ such that $lh = g$ (see [91, p. 12]) and with mappings a, l and r the same as in Mod-R.

If H is a Hopf algebra over a field K, the category of all representations of H is monoidal (cf. [24, Example 5.1.4]).

Numerous examples, different form the above ones, can be found in [24] and [118].

A *Grothendieck category*, i. e., a cocomplete abelian category with exact direct limits and with one generator (see [105, Chapter V]) is called *commutative* if it has a generator U such that its endomorphism ring, $R = End_{\mathcal{C}}(U)$, is commutative. Such a generator is called *endocommutative.* By Gabriel-Popescu theorem, these categories are just the categories equivalent to quotient categories of a category of modules with respect to hereditary torsion theories over commutative rings, i. e., the equivalent ones to a category of the form Mod-(R, τ), of the τ-injective and τ-torsion-free R-modules (cf. [105, Chapter X]). Therefore, they preserve many properties of categories of modules over commutative rings. They have been studied in [7] and their references. By Gabriel-Popescu theorem, there is an equivalence between the categories $\mathcal{C}$ and Mod-(R, τ), where $R = End_{\mathcal{C}}(U)$. Consider the adjoint functors $i : \mathcal{C} \longrightarrow$ Mod-R and a :Mod-$R \longrightarrow \mathcal{C}$, according to Stenström's book notation (cf. [105]), the functor $T : \mathcal{C} \to$Mod-R, given by $T(C) = Hom_{\mathcal{C}}(U, C)$ and the functor $T' : \mathcal{C} \to$ Mod-(R, τ), which is an equivalence of categories.

Lemma 1.2.20 *The category Mod-(R, τ), R being a commutative ring, is a monoidal category.*

Proof: The proof can be seen in [71, Proposition 3.2] or in [60, Proposition 1.2.2]. Only for better illustration, say that $M \otimes N = a(M \otimes_R N)$, the unit is $a(R)$, l_M and r_M are the induced ones by the product of M and R. The tensor product of morphisms is defined by means of $f \otimes g = a(f \otimes_R g)$. Finally, $a_{\{L,M,N\}}$ is the induced one by the tensor $\otimes_R$ and its associativity. ■

Theorem 1.2.21 *If $\mathcal{C}$ is a commutative Grothendieck category, then there exists a monoidal category of the form $(\mathcal{C}, \otimes_{\mathcal{C}}, U, a, l, r)$, where U is an endo-commutative generator of $\mathcal{C}$.*

Proof: Let τ be the torsion theory which appears by applying Gabriel-Popescu theorem to $\mathcal{C}$. The tensor product making Mod-(R, τ) into a monoidal category will be denoted by $\otimes$ (simply). Then, for any objects A, B in $\mathcal{C}$, it is defined $A \otimes_{\mathcal{C}} B = G(T'(A) \otimes T'(B))$, where G is the equivalence of categories inverse of T'. The natural isomorphism between $(A \otimes_{\mathcal{C}} B) \otimes_{\mathcal{C}} C$ and $A \otimes_{\mathcal{C}} (B \otimes_{\mathcal{C}} C)$ will be denoted $a_{\{A,B,C\}}$. The construction of these isomorphisms is because

$$
\begin{aligned}
(A \otimes_{\mathcal{C}} B) \otimes_{\mathcal{C}} C &= G(T'(A \otimes_{\mathcal{C}} B) \otimes T'(C)) \\
&= G(T'G(T'(A) \otimes T'(B)) \otimes T'(C)) \\
&= G((T'(A) \otimes T'(B)) \otimes T'(C))
\end{aligned}
$$

and therefore, $a_{\{A,B,C\}}$ is defined as

$$
G[(1_{T'(A)} \otimes \psi^{-1}_{T'(B) \otimes T'(C)}) \circ a_{\{T'(A),T'(B),T'(C)\}} \circ (\psi_{\{T'(A) \otimes T'(B)\}} \otimes 1_{T'(C)})]
$$

where ψ_M represents the natural isomorphism between $T'G(M)$ and M. Let $l_A = \eta_A \circ G(l_{T'(A)})$ and $r_A = \eta_A \circ G(r_{T'(A)})$, η_A being the natural isomorphism between $GT'(A)$ and A. The tensor product of morphisms is defined as $f \otimes_{\mathcal{C}} g = G(T'(f) \otimes T'(g))$.

Now, we point out that the category Mod-(R, τ), R being a commutative ring, is a monoidal category. The proof can be seen in [71, Proposition 3.2] or in [60, Proposition 1.2.2]. The rest of the proof is easy but tedious and for this reason it is missed. ■

The following result shows that, because of the form in which the tensor product has been defined, the equivalence T' is a monoidal equivalence.

Theorem 1.2.22 *Given a commutative Grothendieck category $\mathcal{C}$, equivalent by means of the functor T' to Mod-(R, τ), T' is a monoidal equivalence between $\mathcal{C}$ (by using the above defined tensor product) and Mod-(R, τ).*

Proof: The isomorphism Φ_0 is $1_{T'(U)}$ and $\Phi_{2,A,B}$ is $\psi^{-1}_{T'(A) \otimes T'(B)}$. Check that the corresponding properties are verified. Then,

$$
\Phi_{2,A,B \otimes_{\mathcal{C}} C} \circ (1_{FA} \otimes \Phi_{2,B,C}) \circ a_{\{FA,FB,FC\}} =
$$

$$\psi^{-1}_{T'(A)\otimes T'(B\otimes_C C)} \circ \left(1_{T'(A)} \otimes \psi^{-1}_{T'(B)\otimes T'(C)}\right) \circ a_{T'(A),T'(B),T'(C)\}}.$$

It holds,

$$F(a_{\{A,B,C\}}) \circ \Phi_{2,A\otimes_C B,C} \circ (\Phi_{2,A,B} \otimes 1_{T'(C)}) =$$
$$T'(a_{\{A,B,C\}}) \circ \psi^{-1}_{T'(A\otimes_C B)\otimes T'(C)} \circ (\psi^{-1}_{T'(A)\otimes T'(B)} \otimes 1_{T'(C)}) =$$
$$\psi^{-1}_{T'(A)\otimes T'(B\otimes_C C)} \circ \left(1_{T'(A)} \otimes \psi^{-1}_{T'(B)\otimes T'(C)}\right) \circ a_{\{T'(A),T'(B),T'(C)\}} \circ$$
$$(\psi_{T'(A)\otimes T'(B)} \otimes 1_{T'(C)}) \circ (\psi^{-1}_{T'(A)\otimes T'(B)} \otimes 1_{T'(C)}),$$

by naturality, and this expression is equal to the above one by simplifying the last two parenthesis. Thus **fu1** is verified.

In our case,

$$T'(r_A) \circ \Phi_{2,A,I} \circ (1_{T'(A)} \otimes \Phi_0) = T'(r_A) \circ \psi^{-1}_{T'(A)\otimes T'(U)} \circ (1_{T'(A)} \otimes 1_{T'(U)})$$
$$= T'(r_A) \circ \psi^{-1}_{T'(A)\otimes T'(U)}.$$

By applying T' to the equality $r_A = \eta_A \circ G(r_{T'(A)})$, it follows that

$$T'(r_A) = T'(\eta_A) \circ T'G(r_{T'(A)}) = \psi_{T'(A)} \circ T'G(r_{T'(A)}).$$

By naturality, the second member of the last equality is

$$r_{T'(A)} \circ \psi_{T'(A)\otimes T'(U)}.$$

By substituting,

$$T'(r_A) \circ \Phi_{2,A,I} \circ (1_{T'(A)} \otimes \Phi_0) = r_{T'(A)} \circ \psi_{T'(A)\otimes T'(U)} \psi^{-1}_{T'(A)\otimes T'(U)} = r_{T'(A)}$$

and therefore, it is verified property **fu2**. Proof for **fu3** is totally analogous and it is omitted. ■

Chapter 2

Multiplication objects in monoidal categories

The concepts of multiplication module and multiplication module relative to a torsion theory (cf. [36] and [37]) have been already studied from a categorical point of view in [38], where the concept of multiplication object in a commutative Grothendieck category is established. However, the natural definition of multiplication graded module, also investigated in [38, Section 5], is not the particularization of the definition of multiplication object in the category gr-R of graded right R-modules and degree preserving homomorphisms. This problem is solved in this work where monoidal categories turn out the ideal setting to research into all these concepts, included that of multiplication graded module. The aim of this chapter is to study multiplication objects in monoidal categories and some particular cases such as the category of right C-comodules, C being a cocommutative coalgebra or the category of Doi-Koppinen modules. The chapter is organized as follows:

In Section 1, some general properties and examples of multiplication objects are exposed and obtained. The study is illustrated with some special cases such as the aforementioned categories. In Section 2, we investigate the case of commutative Grothendieck categories giving another equivalent definition of multiplication object by using any endocommutative generator and Gabriel-Popescu theorem. We apply the result to the categories $\sigma[M]$ and Mod-(R, τ). Section 3 is devoted to exemplify the theory by studying the concepts of multiplication object for comodules, Doi-Koppinen modules or graded modules. In Section 4, we obtain some results on endomorphisms of multiplication objects in braided monoidal categories. We have proved that the endomorphism ring of an indecomposable multiplication object which verifies descendent chain condition on multiplication subobjects is local in any braided monoidal category.

2.1 Properties of multiplication objects

We find some examples of multiplication objects in any monoidal category such as the unit object, simple objects and homomorphic images of multiplication objects.

Throughout this section $(\mathcal{C}, \otimes_{\mathcal{C}}, I, a, l, r)$ (or simply $\mathcal{C}$) stands for a monoidal category. The tensor product $\otimes_{\mathcal{C}}$ will be substituted by $\otimes$ when confusion is not possible.

Definition 2.1.1 *An object X of $\mathcal{C}$ is a multiplication object if every subobject Y of X can be written as $Y = l_X(J \otimes_{\mathcal{C}} X) = r_X(X \otimes_{\mathcal{C}} H)$ for some subobjects H, J of I.*

In the proof of most of results, it is verified only the condition using r_X because with l_X, it is similar. Let us give some examples of multiplication objects.

Proposition 2.1.2 *The unit of every monoidal category is a multiplication object.*

Proof: Let $(\mathcal{C}, \otimes_{\mathcal{C}}, I, a, l, r)$ be a monoidal category and let J be a subobject of I. Since l_J and r_J are isomorphisms, it holds $J = l_J(I \otimes_{\mathcal{C}} J) = r_J(J \otimes_{\mathcal{C}} I)$. Let $i : J \to I$ be the inclusion morphism. By property (2) of monoidal categories, $i \circ l_J = l_I \circ (1_I \otimes_{\mathcal{C}} i)$. Then, $J = i(J) = (i \circ l_J)(I \otimes_{\mathcal{C}} J) = (l_I \circ (1_I \otimes_{\mathcal{C}} i))(I \otimes_{\mathcal{C}} J) = l_I(I \otimes_{\mathcal{C}} J)$. The reasoning for r_I is exactly the same. ∎

Recall that an object S is called *simple* if it has no subobjects except 0 (if the category has zero object) and itself.

Proposition 2.1.3 *Every simple object of a monoidal category is a multiplication object.*

Proof: Let S be a simple object. Consider the inclusion morphism $i : 0 \to I$. By property (2) of monoidal categories, it holds $i \circ l_0 = l_S \circ (i \otimes_{\mathcal{C}} 1_S)$. Thus $0 = i(0) = i[l_0(0 \otimes_{\mathcal{C}} S)] = l_S[(i \otimes_{\mathcal{C}} 1_S)(0 \otimes_{\mathcal{C}} S)] = l_S(0 \otimes_{\mathcal{C}} S)$. Analogously, it is proved that $0 = r_S(S \otimes_{\mathcal{C}} 0)$. On the other hand, since l_S and r_S are isomorphisms, it holds $S = l_S(I \otimes_{\mathcal{C}} S) = r_S(S \otimes_{\mathcal{C}} I)$. ∎

Proposition 2.1.4 *Let X be a multiplication object in the monoidal category $(\mathcal{C}, \otimes, I, a, l, r)$ and let $f : X \to Y$ be a morphism in $\mathcal{C}$. Then, $f(X)$ is a multiplication object in $\mathcal{C}$.*

Proof: Every subobject of $f(X)$ is of the form $f(Y)$, Y being a subobject of X. Since X is multiplication, there exists $A \leq I$ such that $Y = l_X(A \otimes_{\mathcal{C}} X)$. By applying f, it holds $f(Y) = (f \circ l_X)(A \otimes_{\mathcal{C}} X)$. By property (2) of monoidal categories, $f \circ l_X = l_{f(X)} \circ (1_I \otimes_{\mathcal{C}} f)$. Therefore,

$$f(Y) = l_{f(X)}[(1_I \otimes_{\mathcal{C}} f)(A \otimes_{\mathcal{C}} X)] = l_{f(X)}(A \otimes_{\mathcal{C}} f(X)).$$

It follows that $f(X)$ is a multiplication object. ■

Remark 2.1.5 Since a quotient object of X is the image of X by the canonical projection, we have that *every quotient object of a multiplication object is also multiplication.*

The following theorem shows that multiplication objects are preserved by monoidal equivalences.

Theorem 2.1.6 *Let $F : \mathcal{C} \to \mathcal{D}$ be a monoidal equivalence between two monoidal categories. Then, X is a multiplication object in $\mathcal{C}$ if and only if $F(X)$ is multiplication in $\mathcal{D}$.*

Proof: Since F is an equivalence of categories, every subobject of $F(X)$ is of the form $F(Y)$, where Y is an subobject of X. As X is multiplication, $Y = l_X(J \otimes X) = r_X(X \otimes H)$, for some subobjects J, H of I. By applying F, we obtain that $F(Y) = F(l_X)(F(J \otimes X)) = F(r_X)(F(X \otimes H))$. On the other hand, $F(J \otimes X) = \Phi_{2,J,X}(FJ \otimes FX) = \Phi_{2,J,X} \circ (\Phi_0 \otimes 1_{FX})(J \otimes FX)$. Therefore, $F(Y) = [F(l_X) \circ \Phi_{2,J,X} \circ (\Phi_0 \otimes 1_{FX})](J \otimes FX)$. By **fu3**, $F(Y) = l_{FX}(J \otimes FX)$. The other condition is similarly proved. ■

Remark. Because of the proof of Theorem 2.1.6, it is clear that *every dense monoidal funtor preserves multiplication objects.*

2.2 Commutative Grothendieck categories

We give another equivalent categorical definition of multiplication object for this kind of categories which generalizes that of multiplication module. It is proved that multiplication objects are preserved by equivalences between commutative Grothendieck categories. We show that it it is possible to define multiplication module for non-commutative rings and this concept is Morita invariant. We obtain some applications in the case of the category $\sigma[M]$. It is proved that M is a τ-multiplication R-module if and only if M_τ, the module of quotients with respect to τ is a multiplication object in Mod-(R, τ).

Let $\mathcal{C}$ be a commutative Grothendieck category with an endocommutative generator, U, and $R = End_{\mathcal{C}}(U)$. Let X be an object of $\mathcal{C}$. Given a morphism $r \in R$, since $\mathcal{C}$ is commutative, it is possible to consider, for X, the morphism induced by the multiplication by r, i. e., $r^* : T(X) \to T(X)$, i. e., $r^*(\alpha) = \alpha \circ r$ for every morphism α from U in X. Since T is a full functor, there exists an endomorphism of X, r^{**}, such that $T(r^{**}) = r^*$, i. e., $r^{**} \circ \alpha = \alpha \circ r$ for every morphism $\alpha : U \to X$ (since T is faithful, this assignment is unique). By taking into account all these considerations, we give the next definition.

Definition 2.2.1 *An object X of $\mathcal{C}$ is called a U-multiplication object if for every subobject Y of X there exists a family $\{r_i\}_{i \in I}$ of endomorphisms of U such that $Y = \sum_{i \in I} r_i^{**}(X)$.*

We shall prove that the above definition does not depend on the endocommutative generator.

Proposition 2.2.2 *If X is a U-multiplication object of $\mathcal{C}$ and V is another generator having $End_{\mathcal{C}}(V)$ commutative, then X is also a V-multiplication object.*

Proof: The situation will be as above, a functor $T : \mathcal{C} \to$ Mod-R, R being the endomorphism ring of U (recall that $T = Hom_{\mathcal{C}}(U, -)$). The corresponding functor and the corresponding ring for V will be denoted by L and S respectively. Let Y be a subobject of the U-multiplication object X. Then, there exists a family of endomorphisms of U, $\{r_i\}_{i\in I}$ such that $Y = \sum_{i\in I} r_i^{**}(X)$, which is obtained as it has been explained before. Therefore,

$$T(r_i^{**})(\alpha) = r_i^{**} \circ \alpha = \alpha \circ r_i \qquad (1)$$

for every morphism $\alpha : U \to X$. Since V is an object of $\mathcal{C}$, there will exist endomorphisms r_i^{vv} $(i \in I)$ verifying

$$T(r_i^{vv})(\beta) = r_i^{vv} \circ \beta = \beta \circ r_i \qquad (2)$$

for every morphism $\beta : U \to V$. On the other hand, and following again the construction appeared in the Gabriel-Popescu theorem, $s_i^{**} : X \to X$ are defined verifying that $L(s_i^{**})(\gamma) = s_i^{**} \circ \gamma = \gamma \circ r_i^{vv}$ for every morphism $\gamma : V \to X$.

To continue the proof, we will prove that $T(r_i^{**} \circ \gamma) = T(\gamma \circ r_i^{vv})$ for every morphism $\gamma : V \to X$. In fact, let $\beta : U \to V$. By using (1),

$$T(r_i^{**} \circ \gamma)(\beta) = r_i^{**} \circ \gamma \circ \beta = \gamma \circ \beta \circ r_i.$$

By using (2),

$$T(\gamma \circ r_i^{vv})(\beta) = \gamma \circ r_i^{vv} \circ \beta = \gamma \circ \beta \circ r_i.$$

As T is faithful, $r_i^{**} \circ \gamma = \gamma \circ r_i^{vv}$, i. e., $L(r_i^{**}) = L(s_i^{**})$. Because L is faithful, $r_i^{**} = s_i^{**}$ and, consequently, $Y = \sum_{i\in I} s_i^{**}(X)$. ■

As a result, it is possible to give the following definition which generalizes that of multiplication module. Let R be a commutative ring with an identity element. Recall that an R-module M is called *multiplication* if for every submodule N there exists an ideal I of R such that $N = M.I$.

Definition 2.2.3 *An object X of a commutative Grothendieck category $\mathcal{C}$ is said to be a multiplication object if it is U-multiplication for some endocommutative generator U of $\mathcal{C}$.*

Proposition 2.2.4 *Multiplication objects in Mod-R (R being commutative) are just all the multiplication R-modules.*

Proof: Suppose that X is an R-multiplication object in Mod-R. Let Y be an R-submodule of X. Then, it is also a subobject and by hypothesis, there exist $\{r_i\}_{i\in I}$ endomorphisms of R such that $Y = \sum_{i\in I} r_i^{**}(X)$ verifying $r_i^{**} \circ \alpha = \alpha \circ r_i$ for every $i \in I$ and for every morphism $\alpha : R \to X$. R is identified (multiplicatively) with $Hom_R(R, R)$, i. e., an endomorphism f of R is substituted by $f(1)$. Then, $r_i^{**}(X) = X.r_i(1)$ and hence, $Y = \sum_{i\in I} X.r_i(1)$. Calling $H = \{r_i(1)\}_{i\in I}$, it is clear that $Y = X.H$. Thus, $H \subseteq (Y : X)$. By multiplying by X, it is obtained $Y = X.H \subseteq X.(Y : X)$ and, as the other inclusion always holds, it follows that $Y = X.(Y : X)$.

Conversely, suppose that X is a multiplication R-module. Let Y be a subobject of X, i. e., an R-submodule. By hypothesis, $Y = X.I$ for some ideal I of R. It suffices to consider the family $\{r_i\}_{i\in I}$ of endomorphisms of R such that $r_i(1) = i$. It is clear $Y = \sum_{i\in I} r_i^{**}(X)$. ■

Remark. According to the terminology of [117], X is a multiplication object in Mod-R when it generates (multiplicatively) all its subobjects. In particular, the multiplication objects in Mod-R are *self-generators.* Then, it possible to think whether being a self-generator jointly with another property characterizes the fact of being a multiplication module. The following result answers this question. Denote by $S = End(M_R)$ the endomorphism ring of M and by $B = BiEnd(M_R) = End(_SM)$ the biendomorphism ring of M. Recall that an R-module is called *balanced* if every element of B is obtained by multiplying by an element $r \in R$ (cf. [12, p. 61]).

Definition 2.2.5 *An R-module M is said to be a biself-generator if it is a self-generator that generates all its submodules by means of biendomorphisms of M.*

Proposition 2.2.6 *Let M be a finitely generated module. Then, M is a multiplication module if and only if it is a biself-generator and balanced.*

Proof: Every multiplication module is self-generator as it has been exposed in the above remark. If, moreover, M is finitely generated, then its endomorphism ring $End(M_R)$, is isomorphic to the ring $R/ann(M)$ (cf. [83]). As $B \subseteq End(M_R)$, every element of B is obtained multiplying by some $r \in R$.

Now, suppose that the right side of the equivalence is true. Let N be a submodule of M. Since M is a biself-generator, there exist endomorphisms of M, φ_i $(i \in I)$ such that $N = \sum_{i \in I} \varphi_i(M)$. Because M is balanced, every φ_i is obtained multiplying by some $r_i \in I$. By Proposition 2.2.4, M is multiplication as an R-module. ∎

The following result shows that equivalences of commutative Grothendieck categories preserve multiplication objects.

Proposition 2.2.7 *If $F : \mathcal{C} \to \mathcal{D}$ is an equivalence between two commutative Grothendieck categories and X is a multiplication object in $\mathcal{C}$, then $F(X)$ is a multiplication object in $\mathcal{D}$.*

Proof: Let X be a multiplication object of $\mathcal{C}$. Since F is dense, any subobject of $F(X)$ is of the form $F(Y)$, Y being a subobject of X. By hypothesis, there exist endomorphisms r_i $(i \in I)$ of an endocommutative generator U of $\mathcal{C}$, such that $Y = \sum_{i \in I} r_i^{**}(X)$ where r_i^{**} is constructed as above. Since F is an equivalence, $F(U)$ is an endocommutative generator of $\mathcal{D}$. By the same reason, $F(r_i^{**})$ is constructed from $F(r_i)$ analogously to the construction of r_i^{**} from r_i. Moreover, by applying F to the above formula, taking into account that F is an equivalence, it is obtained that $F(Y) = \sum_{i \in I} F(r_i^{**})(F(X))$. ■

In the case of categories of modules, we can say that being multiplication module or ring is Morita invariant, as we shall check in the next result. Now, rings are not necessarily commutative and then, we say that a module M is *multiplication* if any submodule N can be written as $N = MI$ for some ideal I of the ring R. This definition is due to Ukegawa (see [108]).

One of the possible definitions of Morita equivalence is the following: rings R and S are Morita equivalent if there exist bimodules ${}_SP_R$ and ${}_RQ_S$ and surjective morphisms of bimodules $\alpha : P \otimes_R Q \to_S S_S$ and $\beta : Q \otimes_S P \to {}_RR_R$ such that $\alpha(x \otimes f)y = x\beta(f \otimes y)$ and $f\alpha(x \otimes g) = \beta(f \otimes x)g$ for any $x, y \in P$ and $f, g \in Q$ (cf. [12, Exercise 22.7]). In [12, Exercise 22.8], is indicated that given any ideals I of R and I' of S, the maps $\gamma(I) = \beta(PI \otimes Q)$

and $\delta(I') = \beta(QI' \otimes P)$ define isomorphisms, mutually inverse, between the lattices of ideals of R and S. Moreover, $\gamma(IJ) = \gamma(I)\gamma(J)$ for any I, J ideals of R. By taking into account these considerations, it is easy to prove the following results that shows that being multiplication module or ring is Morita invariant.

Theorem 2.2.8 *Assume that $F : Mod - R \to Mod - S$ is a Morita equivalence. Then, the R-module M is multiplication if and only if the S-module $F(M)$ is multiplication.*

Proof: By using the converse equivalence, it suffices to prove that if M is multiplication, so is $F(M)$. We know that there exist maps α, β, γ and δ, such as the defined above. Since F is dense, every subobject of $F(M)$ is of the form $F(N)$, N being an R-submodule of M. By hypothesis, there exists an ideal I of R such that $N = MI$. It is well-known that F can be taken as $- \otimes_R Q$ (cf. [105, Corollary IV.10.2]. Then, we have $F(N) = F(MI) = MI \otimes_R Q = M \otimes_R IQ$. As a consequence of the first condition of Morita over the maps α and β, it holds $Q\alpha(P \otimes_R IQ) = \beta(Q \otimes_S P)IQ = IQ$ as β is surjective. Therefore, $F(N) = M \otimes_R Q\alpha(P \otimes_R IQ) = (M \otimes_R Q)\alpha(P \otimes_R IQ) = F(M)\alpha(P \otimes_R IQ)$. Since $\alpha(P \otimes_R IQ)$ is an ideal of S, it follows that $F(M)$ is multiplication as an S-module. ■

Corollary 2.2.9 *Suppose that R and S are two Morita equivalent rings. Then, R is multiplication if and only if S is multiplication.*

Proof: First, notice that, by definition of equivalence, it suffices to prove that if R is multiplication,then any ideal of S is multiplication as well. Then, every ideal of S is of the form $\gamma(I)$ for some ideal I of R. Suppose that

$\gamma(J) \subseteq \gamma(I)$ for ideals I, J of R. By applying δ, we obtain that $I \subseteq J$. By hypothesis, $J = AI$ for some ideal A of R. Now, we apply γ and we obtain $\gamma(J) = \gamma(AI) = \gamma(A)\gamma(I)$, which proves that S is multiplication. ■

Let R be a commutative ring. Then, recall that an R-module M is *invertible* if there exists $N \in$ Mod-R such that $M \otimes_R N \cong R$ (cf. [114, p. 8]).

Corollary 2.2.10 *Let P be an invertible R-module. Then, if M is a multiplication R-module, $M \otimes_R P$ is multiplication as well.*

Proof: Every invertible R-module is finitely generated, projective and a generator of Mod-R. Therefore, it defines a Morita autoequivalence of Mod-R because P is invertible implies that $End(P_R) \cong R$. Now, it suffices to apply Theorem 2.2.8. ■

Another Grothendieck category where is interesting to study multiplication objects is $\sigma[M]$. Recall some definitions and results which can be found, for example, in [117].

An R-module N is *M-generated* if there is an epimorphism from a direct sum of copies of M into N. Given an R-module M, an R-module N is said to be *subgenerated* by M if it is isomorphic to a submodule of a M-generated R-module. The category $\sigma[M]$ is the full subcategory of Mod-R whose objects are the R-modules subgenerated by M.

In some cases, studying multiplication objects in $\sigma[M]$ is very easy as we can check in the next results.

Proposition 2.2.11 *If M is finitely generated or R is artinian, then M is a multiplication object in $\sigma[M]$ if and only if it is multiplication as an $R/ann(M)$-module.*

Proof: By [117, Propositions 15.4 and 31.5], $\sigma[M] = R/ann(M) - Mod$. Now, it suffices to apply Proposition 2.2.7. ■

It is well-known that $\sigma[M]$ is a Grothendieck category and the direct sum of cyclic submodules of $M^{(\mathbb{N})}$ is a generator, which will be denoted by U. We suppose that this generator is endocommutative and therefore, the Grothendieck category is commutative.

Proposition 2.2.12 *Every multiplication module in $\sigma[M]$ is a multiplication object in $\sigma[M]$.*

Proof: Let N be a multiplication module belonging to $\sigma[M]$. Let N' a subobject of M in $\sigma[M]$, i. e., a submodule of N. By hypothesis, there exists an ideal I of R such that $N' = N.I$. In terms of morphisms, there exist families of morphisms $r_i : R \to R$ and $r_i^{**} : N \to N$ $(i \in I)$ such that $N' = \sum_{i \in I} r_i^{**}(N)$ with $r_i^{**} o \alpha = \alpha o r_i$ for every morphism $\alpha : R \to N$. To be exact, $r_i^{**}(n) = n.r_i(1)$ for every $n \in M$, $i \in I$. For every cyclic submodule Rx of $M^{(\mathbb{N})}$, it is considered the endomorphism induced by multiplying by r_i and by definition of direct sum , it is induced the corresponding endomorphism $\varphi_i : U \to U$. To finish, check that $r_i^{**} \circ \beta = \beta \circ \varphi_i$ for every morphism $\beta : U \to N$. In fact, if $x = x_{i_1} r_{i_1} + \cdots + ... + x_{i_n} r_{i_n} \in U$, then $(\beta \circ \varphi_i)(x) = \beta(x_{i_1} r_{i_1} r_i(1) + \cdots + x_{i_n} r_{i_n} r_i(1)) = \beta(x) r_i(1) = r_i^{**}(\beta(x)) = (r_i^{**} \circ \beta)(x)$. ■

As above, we denote by $S = End(M_R)$ the endomorphism ring of M and by $B = BiEnd(M_R) = End({}_S M)$ the biendomorphism ring of M. The following proposition is a partial inverse of Proposition 2.2.12 for the specific R-module M.

Proposition 2.2.13 *Assume that M is a balanced module and an endocommutative generator of $\sigma[M]$. In these circumstances, if M is a multiplication object in $\sigma[M]$, then M is a multiplication module.*

Proof: Let N be a submodule of M. Since M is multiplication in $\sigma[M]$, there exist families of morphisms $\varphi_i, \varphi_i^{**} : M \to M$ $(i \in A)$ such that $N = \sum_{i \in A} \varphi_i^{**}(M)$ and moreover, $\varphi_i^{**} \circ \beta = \beta \circ \varphi_i$ for every $i \in A$ and for every endomorphism β of M. For $\beta = 1_M$, it is deduced that $\varphi_i^{**} = \varphi_i$. Therefore, the stated condition remains as $\varphi_i \circ \beta = \beta \circ \varphi_i$. Hence, $\varphi_i \in End({}_SM)$ and since M is balanced, $\varphi_i(m) = mr_i$ for some $r_i \in R$. Consider $r_i : R \to R$ defined by $r_i(1) = r_i$ for every $i \in A$. It is immediate that $\varphi_i \circ \alpha = \alpha \circ \varphi_i$. In fact, $(\varphi_i \circ \alpha)(r) = \alpha(r)r_i = (\alpha \circ r_i)(r)$ for every $r \in R$. ■

Remark. It is convenient to emphasize, between the particular cases to apply the theorem, that of M being finitely generated and generator of $\sigma[M]$. In fact, by the Theorem of Density (cf. [117]), it is obvious that, in these circumstances, M is balanced.

An R-module N is called *M-projective* if for every quotient module X of M the canonical morphisms $Hom_R(N, M) \to Hom_R(N, X)$ are epimorphisms. M is called *quasiprojective* if it is M-projective. M is called Σ-quasiprojective if every direct sum of copies of M is quasiprojective. An R-module N is called *M-faithful* if is torsion-free for the torsion theory defined on $\sigma[M]$ whose torsion class is $\{L \in \sigma[M]; Hom_R(M, L) = 0\}$ (cf. [48, p. 89]). Let $S = End(M)$. Consider τ the hereditary torsion theory with corresponding Gabriel filter $\mathcal{F} = \{I \leq S; IM = M\}$. With these concepts, it is possible to express the following result.

Proposition 2.2.14 *If M is Σ-quasiprojective, then an M-generated M-faithful module N is a multiplication object in $\sigma[M]$ if and only if the S-module $Hom_R(M,N)$ is τ-multiplication. If, moreover, M is finitely generated, then N is a multiplication object in $\sigma[M]$ if and only if $Hom_R(M,N)$ is a multiplication S-module.*

Proof: Since M is Σ-quasiprojective, by [48, Theorem 1.3], Mod-(S,τ) is equivalent to certain quotient category of $\sigma[M]$, which is equivalent to the subcategory of M-generated M-faithful modules by [55, p. 113]. As an immediate consequence the first part of proposition is proved. In the finitely generated case, the equivalence holds between the mentioned subcategory of $\sigma[M]$ and Mod-S. By Proposition 2.2.7, the thesis follows. ■

Let R be a commutative ring, τ a hereditary torsion theory over R and M an R-module. The functor a :Mod-R →Mod-(R,τ) maps the R-module M into its module of quotients with respect to τ, the R-module M_τ (see [105] for further information).

Let R be a commutative ring and let τ be a hereditary torsion theory on R, whose corresponding Gabriel filter will be denoted by $\mathcal{F}$. M represents a right R-module. The functor a applies M to the R-module M_τ. This module is called *module of quotients of M with respect to τ* and it is the direct limit, taken in the set of ideals I of the filter with the order inverse to the inclusion, of the system consisting of the sets $Hom_R(I, M/\tau(M))$. The following theorem reestablishes the concept of multiplication module relative to a torsion theory (cf. [36]) inside a more categorical setting. First, we have to obtain some previous results.

Given an R-module M, we denote $M' = M/\tau(M)$ and $R' = R/\tau(R)$. $\mathcal{F}'$ will be the Gabriel filter consisting of the ideals of the form $I/\tau(R)$ where $I \in \mathcal{F}$ and $\tau(R) \subseteq I$ and τ' will be the corresponding torsion theory on R' (cf. [53]). It is obvious that M' is an R'-module. The proof of th following Lemma is omitted because it is straightforward.

Lemma 2.2.15 *M_τ is a multiplication object in Mod-(R, τ) if and only if so is $M'_{\tau'}$ in Mod-(R', τ').*

Let M be a right R-module and let $N \leq M$. N is said to be a *τ-closed submodule* if $N = Cl_\tau^M(N)$. The submodule $\tau(M) = \{x \in M; \exists I \in \mathcal{F} \mid x.I \subseteq 0\}$ is called *τ-torsion submodule* of M. We need the next Lemmas, whose proof is very easy.

Lemma 2.2.16 *Let L be a submodule of M containing the torsion submodule of M, $\tau(M)$. Then $Cl_{\tau'}^{M'}(L/\tau(M)) = Cl_\tau^M(L)/\tau(M)$.*

Definition 2.2.17 *An R-module M is said to be a τ-multiplication module if for every τ-closed submodule N, there exists an ideal I of R such that $N = Cl_\tau^M(M.I)$.*

The next result tells us that if the module is τ-multiplication, we can choose for every submodule the ideal corresponding to the definition.

Lemma 2.2.18 *Let M be an R-module. Then M is τ-multiplication if and only if $X = Cl_\tau^M(M.(X : M))$ for every τ-closed submodule X of M.*

Proof: Suppose that M is τ-multiplication. Let X a τ-closed submodule of M. There exists an ideal I of R, such that $X = Cl_\tau^M(M.I)$. Hence

$$M.I \subseteq Cl_\tau^M(M.I) = X \Rightarrow I \subseteq (X : M).$$

Multiplying by M, we obtain $M.I \subseteq M.(X : M)$. Taking closures

$$Cl_\tau^M(M.I) = X \subseteq Cl_\tau^M(M.(X : M)).$$

On the other hand,

$$M.(X : M) \subseteq X = Cl_\tau^M(M.I)$$

and taking closures again

$$Cl_\tau^M(M.(X : M)) \subseteq Cl_\tau^M(X) = X.$$

The converse implication is trivial. ■

Lemma 2.2.19 *M' is τ'-multiplication as an R'-module if and only if M is τ-multiplication as an R-module.*

Theorem 2.2.20 *M is a τ-multiplication R-module if and only if $a(M)$ is a multiplication object in Mod-(R, τ).*

Proof: Suppose that M is τ-multiplication. Every subobject of $a(M)$ is of the form $a(N)$ with $N \leq M$. By the supposition, there exists an ideal I of R such that $Cl_\tau^M(M.I) = Cl_\tau^M(N)$. By applying a, it is deduced that $a(N) = a(M.I)$. We will see that for every $r \in R$, it holds $a(M).r = a(M.r)$. In fact, denote by r^* the multiplication by r from M into M (by an abuse of language, we also denote of the same form the endomorphism induced in $M/\tau(M)$ and by φ_r the multiplication by r from $a(M)$ into $a(M)$). Let $x \in a(M)$ and suppose that it is represented by $\alpha : A \to M/\tau(M)$ for some $A \in \mathcal{F}$. Then $\varphi_r(x) = x.r$ is an element of $a(M)$ represented by $r.\alpha$. On the other hand, $a(r^*)(x)$ is represented by $r^* \circ \alpha$ (where, in fact, r^* is the induced one). This morphism applied to an element a in A is $r.\alpha(a)$. By the exactness of a (cf. [105], page 215), $a(N) = a(M.I) = a(M).I$. As above, $a(M).I$ can be regarded as a sum of images of endomorphisms of $a(M)$ obtained from the multiplication by elements of R. Moreover, every endomorphism of R (in particular, the multiplication by an element) induces an endomorphism of R_τ (which will be also obtained by means of multiplication). Notice that R_τ is an endocommutative generator of Mod-(R, τ) because $End_{\mathcal{C}}(R_\tau)$ is $Hom_R(R_\tau, R_\tau) = Hom_{R_\tau}(R_\tau, R_\tau)$ by [105, Corollary[IX,1.10] and this ring is isomorphic to R_τ, which is commutative by [105, Exercise IX,2]. Consequently, we can write $a(N) = \sum_{i\in I} \varphi_{i\in I}(a(M))$ verifying the stated conditions. Conversely, suppose that $a(M)$ is multiplication in Mod-(R, τ). Let $N = Cl_\tau^M(N)$. By hypothesis, there exists $A \leq R_\tau$ such that $a(N) = a(M).A$. It can be chosen $A = (a(N) :_{R_\tau} a(M))$. Firstly, suppose that R is τ-faithful ($\tau(R) = 0$). Since the ring is commutative, by [105, Corollary VI.6.6], the Gabriel filter consists of dense ideals (ideals belonging to the filter of the torsion theory cogenerated by the injective hull of R). By [105, Exercise IX.13], for every element of R_τ there is some ideal in the filter which multiplied by that element is inside R. Therefore, now it is easy to see that given $n \in a(N)$,

$n.I \subseteq a(M).(a(N) :_R a(M))$ for some $I \in \mathcal{F}$. Thus

$$\begin{aligned} a(N) &= Cl_\tau^{a(M)}(a(M).(a(N) :_R a(M))) \\ &= Cl_\tau^{a(M)}(a(M.(a(N) :_R a(M))) \\ &= a(M.(a(N) :_R a(M))). \end{aligned}$$

where the exactness of the functor a has been again used. As $a(L) = a(Cl_\tau^M(L))$ for every submodule L of M, N and $Cl_\tau^M(M.(a(N) :_R a(M)))$ are two τ-closed submodules of M corresponding to the same $a(N)$. This correspondence is bijective, according to [53, Proposition 26.20]. For this reason, $N = Cl_\tau^M(M.(a(N) :_R a(M)))$ and M is τ-multiplication. Thus, the theorem is proved for the τ-faithful case. Now, suppose that R is any commutative ring with 1. By hypothesis and by Lemma 2.2.15, $M'_{\tau'}$ is a multiplication object in Mod-(R', τ'). But R' is τ'-faithful and hence, M' is τ'-multiplication as an R'-module. By Lemma 2.2.19, M is a τ-multiplication R-module. ■

We relate the concepts of multiplication object in Mod-R_τ and in Mod-(R, τ) in the following proposition.

Proposition 2.2.21 *If M_τ is a multiplication object in Mod-R_τ, then $a(M)$ is a multiplication object in Mod-(R, τ).*

Proof: Let $a(N)$ be a subobject of $a(M)$. Then, $N_\tau \leq M_\tau$ as R_τ-modules. By hypothesis, there exists a family φ_i $(i \in I)$ of endomorphisms of R_τ verifying that $N_\tau = \sum_{i \in I} \varphi^{**}(M_\tau)$ where the construction of φ^{**} is the exposed one in Definition 2.2.1. Therefore, $a(N) = \sum_{i \in I} \varphi^{**}(a(M))$. Since R_τ is a generator in Mod-(R, τ), it follows that $a(M) = M_\tau$ is a multiplication object in Mod-(R, τ). ■

The following result shows that the definitions of multiplication object in monoidal categories and in commutative Grothendieck categories coincide.

Theorem 2.2.22 *Let $\mathcal{C}$ be a commutative Grothendieck category with endocommutative generator, U and let $(\mathcal{C}, \otimes_\mathcal{C}, U, a, l, r)$ be the corresponding monoidal category. An object X of $\mathcal{C}$ is multiplication in $\mathcal{C}$ if and only if it is multiplication in $(\mathcal{C}, \otimes_\mathcal{C}, U, a, l, r)$.*

Proof: Let $R = End_\mathcal{C}(U)$. Then, Gabriel-Popescu theorem says that $\mathcal{C}$ is equivalent to (via T') the category Mod-(R, τ). Every object of Mod-(R, τ) is of the form $a(M)$ for some $M \in$ Mod-R. Then, $a(M)$ is multiplication in the monoidal category Mod-(R, τ) if and only if, for every subobject $a(N)$ $(N \leq M)$ of $a(M)$, there exists an ideal A of R such that $a(N) = r_{a(M)}(a(M) \otimes a(A)) = a(M).a(A)$. By following the same steps as in the proof of Theorem 2.2.20, that is equivalent to say that $a(M)$ is a multiplication object in the commutative Grothendieck category Mod-(R, τ). Since T' is an equivalence of commutative Grothendieck categories mapping endocommutative generator to endocommutative generator, by Proposition 2.2.7, the object X of $\mathcal{C}$ is multiplication if an only if so is $T'(X)$ in the commutative Grothendieck category Mod-(R, τ). On the other hand, T' is also a monoidal equivalence. By Theorem 2.1.6, X is multiplication in the monoidal category $\mathcal{C}$ if and only if so is $T'(X)$ in Mod-(R, τ). At the beginning of the proof, we noticed that, in Mod-(R, τ), the two interpretations of multiplication objects are identical and therefore, the proof ends. ■

2.3 Some examples of multiplication objects

As examples, we shall characterize multiplication objects in some particular cases.

Example 1: Graded R-modules. If R is a commutative ring, then a graded R-module M is a *multiplication graded module* (*gr-multiplication*) if every graded submodule N can be written as $N = A.M$ for some ideal (which can be taken graded) A of R. Multiplication graded modules and rings are studied in [38], [39] and [40] but only in some cases the category of graded R-modules is a commutative Grothendieck category. Now, we show that this concept coincide in any case with that of multiplication object in the corresponding monoidal category.

Proposition 2.3.1 *An object M, in the monoidal category gr-R, is multiplication if and only if M is a multiplication graded (gr-multiplication) R-module.*

Proof: The object M is multiplication in gr-R if for every graded submodule, N, of M ($N \leq_{gr} M$), there exists a graded ideal A of R such that $N = r_M(M \otimes A) = l_M(A \otimes M)$. Since $MA = r_M(M \otimes A)$, it is equivalent to say that for every $N \lesssim_{gr} M$, there exists a graded ideal A of R, such that $N = MA$ and this means that M is gr-multiplication. ■

Remark. Given a graded R-module M, the *g-suspension* of M is the R-module $M(g)$ with the graduation defined by $(M(g))_h = M_{hg}$ for every $h \in G$. It is well-known that the functor T_g from gr-R in itself is a monoidal equivalence of categories. As a consequence of Proposition 2.3.1 and Theorem

2.1.6, for any $g \in G$, *the graded R-module M is multiplication if and only if* $M(g)$ *is multiplication.*

If R is a ring graded by G, we can consider the group ring $R[G]$ with the following graduation: $(R[G])_g = \oplus_{h \in G} R_{gh^{-1}} h$ (see [93]). Then, Dade's Theorem give us that the categories Mod-R and gr-$R[G]$ are equivalent. Moreover, this equivalence is clearly monoidal. By Theorem 2.1.6, *R is a multiplication ring if and only if the ring R[G] is gr-multiplication.*

Example 2. Recall from [66, Proposition XV.1.2] that a *quasi-bialgebra* is a K-algebra A with algebra morphisms $\Delta : A \to A \otimes A$ and $\varepsilon : A \to K$ such that the category of left A-modules $(A{-}Mod, \otimes_K, K, a, l, r)$ is monoidal, where $\otimes_K$ is the usual tensor product in K a, l, and r are the usual morphism families. In [97, Theorem 2], it appears the dual concept: a *semi-coalgebra* is a coalgebra with coalgebra morphisms $M : C \otimes C \to C$ and $u : K \to C$ such that the category of right C-comodules $(\mathcal{M}^C, \otimes_K, K, a, l, r)$ is monoidal (where a, l, r are the usual ones).

Proposition 2.3.2 *If A is a quasi-bialgebra, then an A-module M is a multiplication object in* $(A - Mod, \otimes_K, K, a, l, r)$ *if and only if* $M = 0$ *or M is simple. If C is a semi-coalgebra, then a C-comodule M is a multiplication object in* $(\mathcal{M}^C, \otimes_K, K, a, l, r)$ *if and only if* $M = 0$ *or M is a simple comodule, i.e., it has no non-zero proper subcomodule.*

Proof: Suppose that M is a multiplication object in A-Mod. Then for any A-submodule N there exist submodules of K, I and I' such that $N = r_M(M \otimes_K I) = l_M(I' \otimes M)$. But the only possibilities for I and I' are 0 and K. Consequently, M has to be simple. Conversely, every simple module is multiplication by Proposition 2.1.3. The proof for comodules is analogous. ■

Example 3. Consider a family of groups $(G_i)_{i\in\mathbb{N}}$ with $G_0 = \{1\}$ and group morphisms $\rho_{n,m} : G_n \times G_m \to G_{n+m}$ for any $(n,m) \in \mathbb{N} \times \mathbb{N}$. We define $n \otimes m = n + m$ and $g \otimes h = \rho_{n,m}(g,h) \in G_{n+m}$. If $\rho_{n+m,p} \circ (\rho_{n,m} \otimes 1_{G_p}) = \rho_{n,m+p} \circ (1_{G_n} \otimes \rho_{m,p})$ and $\rho_{0,n} = \rho_{n,0} = 1_{G_n}$ for any $n, m, p \in \mathbb{N}$ (after natural identification), then the category $(\mathcal{G}, \otimes, 0, 1, 1, 1)$ is monoidal, where the class of objects of $\mathcal{G}$ is $\mathbb{N}$, the set $Hom_{\mathcal{G}}(i,j)$ is $\emptyset$ if $i \neq j$ and G_i if $i = j$ and the composition in $Hom_{\mathcal{G}}(i,i)$ is the group operation (see [66, XI.3.2]). Since in this category the only subobject of n is itself, *every object is a multiplication object.*

Example 4: Multiplication comodules. For this example (C, Δ, ε) stands for a cocommutative coalgebra over a field K with comultiplication Δ and counit ε. The category of right C-comodules will be denoted by $\mathcal{M}^C$. We follow the notation of [107]. If $M \in \mathcal{M}^C$, then its structure mapping will be $w_M : M \to M \otimes C$. The corresponding structure mapping of M as a left C-comodule will be denoted by w'_M. Recall that given two right C-comodules M and N, the *cotensor product* of M and N, $M\square_C N$, is the kernel of the mapping $\alpha = w_M \otimes 1_N - 1_M \otimes w'_N : M \otimes N \to M \otimes C \otimes N$. We write $i_{M,N}$ for the injection from $M\square_C N$ into $M \otimes N$. By the definition of kernel, there exists a unique morphism $\varphi : M \to M\square_C C$ such that $i_{M,N} \circ \varphi = w_M$. Moreover, it is well-known that it is an isomorphism. The inverse isomorphism of φ will be denoted by r_M. In the same way, it is defined the isomorphism $l_M : C\square_C M \to M$. The isomorphisms

$a_{L,M,N} : (L\Box_C M)\Box_C N \to L\Box_C(M\Box_C N)$ are the induced by the associativity isomorphisms from $(L \otimes M) \otimes N$ into $L \otimes (M \otimes N)$. It is easily seen that $(\mathcal{M}^C, \Box_C, C, a, l, r)$ is a monoidal category. As a consequence, with the above notation and by taking into account the cocommutativity, we are able to give the following definition. Henceforth, for simplicity of notation we write $\Box$ instead of $\Box_C$.

Definition 2.3.3 *A right C-comodule M is called multiplication if for every C-subcomodule N there exists a right coideal D of C such that $r_M(M\Box D) = N$ or equivalently, $M\Box D = w_M(N)$.*

Definition 2.3.4 *A cocommutative coalgebra C is multiplication if every right coideal is multiplication as a C-comodule.*

Any cocommutative coalgebra C is multiplication as a C-comodule by Proposition 2.1.2. A right C-comodule M is called *cyclic* if it is isomorphic to a quotient of the form C/I for some right coideal I of C. From Remark 2.5, every cyclic right comodule is multiplication. A *cosemisimple coalgebra* is a coalgebra which is the sum of its simple subcoalgebras. Then, by Remark 3.7, *every cosemisimple cocommutative coalgebra is a multiplication coalgebra.*

Remark. *There are right comodules which are multiplication in $\mathcal{M}^C$ but they are not multiplication as C^*-modules.*

Let $S = \{c_0, c_1, ...\}$ and C the divided power coalgebra, that is, $C = KS$ with $\Delta(c_i) = \sum_{j=0}^{n} c_j \otimes c_{i-j}$ and $\varepsilon(c_i) = \delta_{0,i}$. C is obviously a cocommutative coalgebra and therefore, C is a multiplication C-comodule. It is well-known that C^* is isomorphic to the power series ring $K[[X]]$. The C^*-modules of C

are of the form 0, C and $\sum_{i=0}^{n} Kc_i$ for some $n \in \mathbb{Z}^+$ (cf. [107, p. 44]). Then, the morphim of C^*-modules $\varphi : M_n = \sum_{i=0}^{n} Kc_i \rightarrow (K[[X]]/(x^{n+1})) = M'_n$ given by

$$\varphi(\sum_{i=0}^{n} k_i c_i) = \sum_{i=0}^{n} k_{n-i} x^i + (x^{n+1})$$

is clearly an isomorphism. Therefore ${}_{C^*}C \cong \cup_{n \in \mathbb{N}} M'_n$ where $M'_n \subseteq M'_{n+1}$ for all $n \in \mathbb{N}$. But this C^*-module is not multiplication as there is no ideal I of $K[[X]]$ such that $C = \frac{K[[X]]}{(X)}.I$. This proves the statement.

Recall from [?] that a C-comodule M is *invertible* if there exists another C-comodule M' and isomorphisms of C-comodules $f : C \rightarrow M \square M'$ and $g : C \rightarrow M' \square M$ such that

a) $(id_M \square g) \circ w_M = a_{\{M,M',M\}} \circ (f \square id_M) \circ w'_M$.

b) $(id_{M'} \square f) \circ w_{M'} = a_{\{M',M,M'\}} \circ (g \square id_{M'}) \circ w'_{M'}$.

The next proposition provides us new examples of multiplication comodules. We use the above notation.

Proposition 2.3.5 *If M is an invertible C-comodule, then M is multiplication in $\mathcal{M}^C$.*

Proof: Let L be a subcomodule of M. We shall prove that $M \square g^{-1}(M' \square L)$ and $W_M(L)$ are equal by checking that their images under the monomorphism $id_M \square g$. On the one hand, $(id_M \square g)(M \square g^{-1}(M' \square L)) = M \square (M' \square L)$.

On the other hand,

$$\begin{aligned}(id_M\Box g)(w_M(L)) &= a_{\{M,M',M\}}\circ(f\Box id_M)(w'_{M'}(L))\\ &= a_{\{M,M',M\}}\circ(f\Box id_M)(C\Box L)\\ &= a_{\{M,M',M\}}(f(C)\Box L)\\ &= a_{\{M,M',M\}}((M\Box M')\Box L)\\ &= M\Box(M'\Box L).\end{aligned}$$

By using b), in the same way, it is proved that $w'_M(L) = f^{-1}(L\Box M')\Box M$. ■

Let $M \in \mathcal{M}^C$ and take $m \in M$. Then, there exists a subcomodule, which is moreover finite-dimensional, of M containing m (see [79, Theorem 5.1.1]). It will be denoted by $< m >$. This notation will be used in the sequel.

Proposition 2.3.6 *A C-comodule M is multiplication if and only if for every $m \in M$ there exists a right coideal I of C such that $w_M(< m >) = M\Box I$.*

Proof: The necessity is clear. Now, suppose that for every $m \in M$, $w_M(< m >) = M\Box I_m$ for some right coideal I_m of C. Let N be a subcomodule of M. Then, $w_M(< n >) = M\Box I_n$ for every $n \in N$. Let us consider $I = \sum_{n\in N} I_n$. Since the sum of subcomodules is a subcomodule, I is a right coideal. We have $w_M(N) = w_M(\sum_{n\in N} < n >) = w_M(\sum_{n\in N}(M\Box I_n)) = w_M(M\Box I)$. ■

Example 5: Doi-Koppinen modules. Let ${}_A\mathcal{M}(H)^C$ be the category of left-right G-Doi-Koppinen modules and A-linear C-colinear homomorphisms, where the triple $G = (H, A, C)$ is a monoidal Doi-Hopf datum over a commutative ring R (see [23] for definitions and further information). By [23, Proposition 2.1], $({}_AM(H)^C, \otimes_R, R, a, l, r)$ is a monoidal category. Given two

Doi-Koppinen modules $N \subseteq M$, we write $(N : M)$ for $\{r \in R; rM \subseteq N\}$. Then, we have a characterization of multiplication object similar to that of multiplication modules.

Proposition 2.3.7 *A Doi-Koppinen module M is multiplication if and only if $N = M(N : M)$ for any Doi-Koppinen submodule N of M.*

Proof: By following the remarks of [102, §1], it suffices to prove that $(N : M)$ is a Doi-Koppinen submodule of R. It is clear that is an ideal of R. Let ε_A be the counit of A, $a \in A$ and $r \in (N : M)$. Then, $a.r.M = \varepsilon_A(a).r.M \subseteq \varepsilon_A(a)N \subseteq N$. It follows that $(N : M)$ is a left A-module. Let w_R the structure map for the comodule R, then $w_R(r) = r \otimes 1$ for every $r \in R$. Since $w_{(N:M)} = w_R/(N : M)$, i.e., the restriction of w_R to $(N : M)$, $(N : M)$ is also a right C-comodule. The compatibility conditions are verified as every element of $(N : M)$ is an element of R. ■

2.4 Endomorphisms of multiplication objects

In the setting of braided monoidal categories, we study properties of the endomorphism ring of multiplication objects. We prove that multiplication objects satisfy a version of Fitting's Lemma, that they are totally invariant and that the endomorphism ring of an indecomposable multiplication object with descendent chain condition on multiplication submodules is local.

Definition 2.4.1 *A monoidal category $(\mathcal{C}, \otimes, I, a, l, r)$ is called braided if there exists a family of natural isomorphisms $c_{A,B} : A \otimes B \to B \otimes A$ verifying the following axioms:*

$(B1)$ $(g \otimes f) \circ c_{A,B} = c_{A',B'} \circ (f \otimes g)$ $\forall f : A \to A',\ g : B \to B'$.

$(B2)$ $a_{\{B,C,A\}} \circ c_{A,B\otimes C} \circ a_{\{A,B,C\}} = (1_B \otimes c_{A,C}) \circ a_{\{B,A,C\}} \circ (c_{A,B} \otimes 1_c)$.

$(B3)$ $a^{-1}_{\{C,A,B\}} \circ c_{A\otimes B,C} \circ a^{-1}_{\{A,B,C\}} = (c_{A,C} \otimes 1_B) \circ a^{-1}_{\{A,C,B\}} \circ (1_A \otimes c_{B,C})$.

When the braided category is strict as a monoidal category and, moreover, $c_{A,I} = c_{I,A} = 1_A$ for all object of $\mathcal{C}$, A, the category $\mathcal{C}$ is a *strict braided category.*

Lemma 2.4.2 *If $(\mathcal{C}, \otimes, I, a, r, l, c)$ is a strict braided category, for every object of $\mathcal{C}$, X and every subobject A of I, we have $r_X(X \otimes A) = l_X(A \otimes X)$ for every A.*

Proof: Notice that $c_{X,I}(X \otimes A) = c_{X,A}(X \otimes A)$. In fact, let i_A the inclusion morphism from A into I. By axiom $(B1)$, $(i_A \otimes 1_X) \circ c_{X,A} = c_{X,I} \circ (1_X \otimes i_A)$. By applying every member to $X \otimes A$, we obtain $[(i_A \otimes 1_X) \circ c_{X,A}](X \otimes A) = c_{X,A}(X \otimes A)$ and $[c_{X,I} \circ (1_X \otimes i_A)](X \otimes A) = c_{X,I}(X \otimes A)$.

Now, we have $r_X(X \otimes A) = (l_x \circ c_{X,I})(X \otimes A) = l_X(c_{X,I}(X \otimes A)) = l_X(c_{X,A}(X \otimes A)) = l_X(A \otimes X)$. ■

Definition 2.4.3 *An object X of a category $\mathcal{C}$ is called totally invariant if $f(Y) \subseteq Y$ for every endomorphism of X, f and for every subobject of X, Y.*

Proposition 2.4.4 *If X is a multiplication object in a braided monoidal category $\mathcal{C}$, then X is totally invariant.*

Proof: It is well-known that every braided monoidal category is equivalent to a strict braided monoidal category. Moreover, since the properties of being totally invariant and being multiplication are preserved by equivalences, it is enough to prove the result for a strict category $\mathcal{C}$. Let $Y \leq X$ and let $f \in End_{\mathcal{C}}(X)$. Since X is multiplication, $Y = r_X(X \otimes A)$ for some $A \leq I$, I being the unit of $\mathcal{C}$. By Proposition 2.1.4, $f(X) = r_X(X \otimes B)$ for some $B \leq I$. By applying f to the first equality,

$$f(Y) = (f \circ r_X)(X \otimes A) = r_X \circ (f \otimes 1_I)(X \otimes A) = r_X(f(X) \otimes A).$$

By substituting the second equality,

$$\begin{aligned} f(Y) &= r_X(r_X(X \otimes B) \otimes A) = r_X((r_X \otimes 1_I)(X \otimes B) \otimes A) \\ &= (r_X \circ r_X)((X \otimes B) \otimes A) = (r_X \circ r_X)(X \otimes (B \otimes A)). \end{aligned}$$

By using that $\mathcal{C}$ is braided and strict,

$$\begin{aligned} f(Y) &= (r_X \circ r_X)((X \otimes A) \otimes B) = r_X(r_X(X \otimes A) \otimes B) \\ &= r_X(Y \otimes B) = r_Y(Y \otimes B) \subseteq B. \end{aligned}$$

■

Proposition 2.4.5 *Let $(\mathcal{C}, \otimes, I, a, l, r, c)$ be a braided monoidal category and let X be a multiplication object. Then, if $f, g \in End_{\mathcal{C}}(X)$, $(f \circ g)(Y) = (g \circ f)(Y)$ for every subobject Y of X.*

Proof: It is clear that it suffices to prove it for a strict category. Thus we suppose that $\mathcal{C}$ is strict. Firstly, we deal with the case $Y = X$. By Proposition 2.1.4, $f(X) = r_X(X \otimes A)$ and $g(X) = r_X(X \otimes B)$ for some subojects A, B of I. By applying g to the first equality, we have $g(f(X)) = (g \circ r_X)(X \otimes A) = r_X \circ (g \otimes 1_X)(X \otimes A) = r_X(g(X) \otimes A) = r_X(r_X(X \otimes B) \otimes A)$. Since $\mathcal{C}$ is strict and braided, $(g \circ f)(X) = r_x^2((X \otimes A) \otimes B)$. In a similar way, we have

$f(g(X)) = f(r_X(X \otimes B)) = r_X(f(X) \otimes B) = r_X^2((X \otimes A) \otimes B)$. Now, we prove it for every $Y \leq X$. Since X is multiplication, $Y = r_X(X \otimes C)$ for some $C \leq I$. Then, $f(Y) = r_X(f(X) \otimes C)$. By applying g, $g(f(Y)) = r_X((g \circ f)(X) \otimes C)$. By the above step, $g(f(Y)) = r_X((f \circ g)(X) \otimes C) = f(g(Y))$. ■

Proposition 2.4.6 *Let X be a multiplication object in a braided monoidal category $\mathcal{C}$ and let $\varphi \in End_{\mathcal{C}}(X)$. Then, φ is an epimorphism if and only if it is an isomorphism.*

Proof: The first step of the proof consists of proving that φ is an epimorphism if and only if $\varphi/Y : Y \to Y$ is an epimorphism for every subobject $Y \leq X$. Notice that by Proposition 2.4.4, φ/Y is defined from Y to Y. Suppose that φ is an epimorphism and let $Y \leq X$. Since X is multiplication, $Y = r_X(X \otimes A)$ for some $A \leq I$. By applying φ, we have $\varphi(Y) = \varphi \circ r_X(X \otimes A) = r_X \circ (\varphi \otimes 1_I)(X \otimes A) = r_X(\varphi(X) \otimes A)$. Since φ is an epimorphism, $\varphi(X) = X$. Therefore, $\varphi(Y) = r_X(X \otimes A) = Y$. The other implication is clear. Now we shall prove that if φ is an epimorphism, then it is a monomorphism. In fact, assume that φ is an epimorphism. By the first step when $Y = 0$, we have $\varphi(0) = 0$. It follows that φ is a monomorphism.

■

In the sequel, we suppose that any monoidal category is abelian. The next result is a kind of Fitting's Lemma.

Proposition 2.4.7 *Let X be a multiplication object in a braided monoidal category $\mathcal{C}$ and let $\varphi \in End_{\mathcal{C}}(X)$ such that $\varphi^n(X) = \varphi^{n+1}(X)$ for some positive integer n. Then, $X = Ker\ \varphi^n \cap \varphi^n(X)$.*

Proof: Take $Y = Ker\ \varphi^n \cap \varphi^n(X)$. By Proposition 2.1.4, $\varphi^n(X)$ is multiplication as well. Thus $Y = r_X(\varphi^n(X) \otimes A)$ for some $A \leq I$. By applying φ^n, we have $\varphi^n(Y) = \varphi^n \circ r_X(\varphi^n(X) \otimes A) = r_X(\varphi^{2n}(X) \otimes A)$. By hypothesis, $\varphi^{2n}(X) = \varphi^n(X)$ and consequently, $\varphi^n(Y) = r_X(\varphi^n(X) \otimes A) = Y = 0$. Moreover, by definition of kernel, $(1_X - \varphi^n)(X) \subseteq Ker\ \varphi^n$. As a result, $X = \varphi^n(X) + (1_X - \varphi^n)(X) \subseteq \varphi^n(X) + Ker\ \varphi^n$. The other inclusion is clear and therefore, we have the equality. ■

Remark. The condition mentioned in Proposition 2.4.7 is obtained when X has descendent chain condition over multiplication objects because it suffices to consider the chain of multiplication objects $X \supseteq \varphi(X) \supseteq \cdots$.

Recall that an object X is called *indecomposable* if $X = X_1 \oplus X_2$ (X_1, X_2 being subobjects of X) implies $X_1 = 0$ or $X_2 = 0$.

Corollary 2.4.8 *Let X be an indecomposable multiplication object in a braided monoidal category and let $\varphi \in End_{\mathcal{C}}(X)$ such that $\varphi^n(X) = \varphi^{n+1}(X)$ for some positive integer n. Then, φ is an automorphism if and only if φ is not nilpotent.*

Proof: If φ is an automorphism, so is φ^n and hence, φ is not nilpotent. Now, suppose that φ is not nilpotent. By Proposition 2.4.7, $X = \varphi^n(X) \oplus Ker\ \varphi^n$. Since X is indecomposable, either $Ker\ \varphi^n = 0$ or $\varphi^n(X) = 0$. If $\varphi^n(X) = 0$, then $\varphi^n = 0$ and φ is nilpotent, a contradiction. Thus $Ker\ \varphi^n = 0$ and hence, φ^n is a monomorphism. As a consequence, φ is a monomorphism and therefore, $X = \varphi^n(X)$. It follows that φ^n, and therefore φ, is an epimorphism. ■

Proposition 2.4.9 *Suppose that X is a multiplication object in a braided monoidal category $\mathcal{C}$ and $\varphi \in End_{\mathcal{C}}(X)$ such that $\varphi^n(X) = \varphi^{n+1}(X)$ for some positive integer n. Then, the following conditions are equivalent:*

1. *φ is a monomorphism.*
2. *φ is an epimorphism.*
3. *φ is an automorphism.*

Proof: Suppose that φ is a monomorphism. Then, φ^n is a monomorphism as well. By hypothesis, $\varphi^n(X) = \varphi^n(\varphi(X))$. It follows that $X = \varphi(X)$ and therefore φ is an epimorphism.

Now, assume that φ is an epimorphism. Then $Ker\ \varphi = r_X(X \otimes A)$ for some $A \leq I$ by hypothesis. By applying φ, we have

$$0 = \varphi \circ r_X(X \otimes A) = r_X(\varphi(X) \otimes A) = r_X(X \otimes A) = Ker\ \varphi.$$

Thus φ is a monomorphism. By Corollary 2.4.8, under these circumstances, 2 is equivalent to 3. ■

Theorem 2.4.10 *If X is an indecomposable multiplication object in a braided monoidal category that verifies D.C.C. on multiplication subobjects, then the ring $End(X)$ is local.*

Proof: Firstly, we shall prove by induction that under our hypothesis, if $g = \varphi_1 + \varphi_2$ is an automorphism of X with $\varphi_i \in End(X)$ $(1 \leq i \leq 2)$, then some φ_i is an automorphism. Since g is an automorphism, there exists g^{-1} and we have $1 = g^{-1} \circ \varphi_1 + g^{-1} \circ \varphi_2$. By Corollary 2.4.8, $g^{-1} \circ \varphi_1$ is an automorphism or nilpotent. If it is nilpotent, then $1 - g^{-1} \circ \varphi_1 = g^{-1}\varphi_2$ is an automorphism because it is invertible.

In order to prove that $End(X)$ is local, we take $f, g \in End(X)$ and we assume that $f+g$ is invertible. We shall check that f or g is not invertible. If $f+g$ is invertible and g is not, there exists $h \in End(X)$ such that $(f+g)\circ h = 1_X$. Since g is not invertible, neither is $g \circ h$. By Corollary 2.4.8, $g \circ h$ is nilpotent. Hence $1 - g \circ h = f \circ h$ is invertible and therefore, f is invertible as well. ■

Chapter 3

Multiplication objects in the category of relative injective relative torsionfree modules

We apply torsion theory techniques to study multiplication objects in the category of τ-injective τ-torsionfree R-modules, Mod-(R, τ). In the first section, an alternative definition of relative multiplication module is given by using semiring theory. Then, several important examples of these concepts are exposed and studied. Finally, some properties over homomorphisms between two multiplication modules are investigated. Section 3.2 is concerning with local properties of relative multiplication modules and rings. The localization is made with respect to a torsion theory (mainly, a semicentered torsion theory) and with respect to τ-closed prime ideals. In Section 3.3, the problem of constructing new relative multiplication modules is tackled, considering different operations between τ-multiplication modules such as sum, intersection, direct sum and product. The study of relative distributive modules and their relationship with τ-multiplication modules is the main aim of Section 3.4. Section 3.5 is concerning with the study of multiplication ideals and rings with respect to a torsion theory and a new version of Mott's Theorem is achieved.

3.1 Alternative definitions

We shall start giving another interpretation, by using semiring and semimodule theory, of multiplication and τ-multiplication module. A large number of examples of τ-multiplication rings, ideals and modules will be exposed. It is outstanding the case of any Krull domain with respect to the canonical torsion theory, i. e., the induced one by the prime ideals of height smaller than or equal to one. Some properties related to homomorphisms will be expressed. Multiplication ideals will be characterized by means of other arithmetical conditions. Finally, several chain condition properties will be shown.

3.1.1 Multiplication elements in semimodules

Apart from the categorical interpretation of multiplication and τ-multiplication module, another definition by using semiring theory can be given.

Semirings appeared for their first time with Vandiver in 1934 in connection to the axiomatization of natural numbers y non negative rational numbers (implicitely, they appeared with Dedekind in 1894). They have turn out important tools in applied mathematics and computer sciences. We shall that they have interest in algebra to study different concepts generalizing those of multiplication module and ring.

A *semiring* is a non empty set R in which two operations, + y ., have been defined satisfying the following conditions:

1. $(R, +)$ is a commutative monoid with an identity element, 0.
2. $(R, .)$ is a semigroup with an identity element, 1.
3. Multiplication is distributive with respect to addition from both sides.
4. $0.r = r.0 = 0$ for all $r \in R$.
5. $1 \neq 0$.

Remark. The last two properties implies that the two operations must be different.

The most outstanding examples are the following: rings, $(\mathbb{N}, +, .)$, the lattice of ideals $\mathcal{L}(R)$ of a ring R with the sum and the product of ideals, the set of fractional ideals, $Frac(R)$, of a ring R, or that of finitely generated fractional ideals, $Frac^*(R)$, with the sum and the product of ideals. Definitions, properties and examples can be found in [54].

Given a semiring R, a *right R-semimodule* is a commutative monoid $(M, +)$ with an additive unit, 0_M, for which it holds a function $M \times R \to M$ called *scalar multiplication* that satisfies the conditions bellow for all elements $r, r' \in R$ and $m, m' \in M$.

1. $m(rr') = (mr)r'$.
2. $(m + m')r = mr + m'r$.

3. $m(r + r') = mr + m'r$.

4. $m1_R = m$.

5. $0_M r = 0_M = m0_R$.

All semimodules will be considered on the right side. As most outstanding examples we shall mention commutative additive monoids which are just $\mathbb{N}$-semimodules and the lattice $\mathcal{L}(M)$ of submodules of an R-module M, which with the sum is a $\mathcal{L}(R)$-semimodule. In order to relate them to multiplication modules, we need to give some definitions.

Definition 3.1.1 *An element m of an R-semimodule M is said to be a multiplication element if for every $n \in M$ such that $n + m = m$, there exists $a \in R$ verifying that $n = ma$.*

Definition 3.1.2 *A semimodule M is called multiplication if every element is multiplication.*

It is obvious that if M is an R-module, then M is multiplication as an R-module if and only if M is a multiplication element in the aforementioned semimodule $\mathcal{L}(M)$. It is also easy to appreciate that the ring R is a multiplication ring if and only if every element of $\mathcal{L}(R)$ is a multiplication element in $\mathcal{L}(R)$. Let τ be a hereditary torsion theory on a ring R. We shall denote by $C_\tau(R)$ and $C_\tau(M)$ the lattices of τ-closed ideals of R and τ-closed submodules of the R-module M respectively. For ideals A, B of R, we define the following operations: $A * B = Cl_\tau^R(A + B)$ and $A \circ B = Cl_\tau^R(A.B)$. Analogous operations, which will be denoted in the same way, are defined for submodules. Scalar multiplication is defined as follows: $N \square A = Cl_\tau^M(NA)$ for every $A \leq R$ and every $N \leq M$.

Proposition 3.1.3 *If R and M are not τ-torsion, then $(C_\tau(R), *, \circ)$ is a semiring and $(C_\tau(M), \square)$ is a $C_\tau(R)$-semimodule.*

Proof: First, we check that $C_\tau(R)$ is a semiring. The first property is verified because $Cl_\tau^R(Cl_\tau^R(A)+Cl_\tau^R(B))$ for any A, B of $C_\tau(R)$ (cf. Proposition 1.1.3, C5) and by taking $Cl_\tau^R(0)$ as the identity element. The second property is true by Proposition 1.1.4 (C8), the ring R being the identity element with respect to $\circ$. The third property is true as a consequence of the two above ones and the distributive property of the product of ideals respect to the sum. The forth one and the fifth one are almost immediate as $Cl_\tau^R(Cl_\tau^R(0).Cl_\tau^R(A)) = Cl_\tau^R(0.A) = Cl_\tau^R(0)$ y $R \neq Cl_\tau^R(0)$.

It is obvious that $(C_\tau(M), \square)$ is a commutative monoid (the identity element is $Cl_\tau^M(0)$). Then, $Cl_\tau^M(NCl_\tau^R(AB)) = Cl_\tau^M(Cl_\tau^M(NA)B)$ is the translation of the first property of semimodule but, by using again the properties of closure operations which appear in Propositions 1.1.3 and 1.1.4, it is easy to check that both expressions are equal to $Cl_\tau^M(NAB)$ for any τ-closed ideals A and B and every τ-closed submodule of M, N. For similar reasons, the second and the third properties hold. The fourth one says that $Cl_\tau^M(NR) = N$, which is evident for a τ-closed submodule N. The fifth one tell us that $Cl_\tau^M(Cl_\tau^M(0)A) = Cl_\tau^M(0) = Cl_\tau^M(N.0)$ and it is trivially true as well. ■

Remark. It has been defined an *τ-multiplication R-module M* as an R-module that verifies that for every τ-closed submodule N, there exists an ideal A of R (which can be always taken τ-closed) such that $N = Cl_\tau^M(MA)$. A *τ-multiplication ring* is a ring that verifies that every ideal is multiplication as a module. Therefore, an R-module M is τ-multiplication if and only if M is a multiplication element in the $\mathcal{C}_\tau(R)$-semimodule $\mathcal{C}_\tau(M)$. Moreover, the ring R is τ-multiplication if and only if every element of $\mathcal{C}_\tau(R)$ is a multiplication element.

3.1.2 Multiplication modules relative to a torsion theory

Some important families of examples of multiplication modules and ring relative to a torsion theory are studied. It is observed that the class of relative multiplication modules is strictly wider than that of classical multiplication modules. Moreover, some technical properties are proved, easily deduced from the definition, relative multiplication ideals are characterized arithmetically and it is proved that it suffices to check the condition stated in the definition of relative multiplication module for τ-cyclic submodules.

If the Gabriel filter is $\mathcal{F}$={ R} then every submodule of a module M is τ-closed. Therefore, M is τ-multiplication if and only if M is multiplication. Therefore, this definition generalizes that of multiplication module given by Barnard in [13], which is the most spread and used. Before this definition, Singh and Mehdi gave another one more demanding definition. Its natural generalization with respect to a torsion theory is the following:

Definition 3.1.4 *An R-module M is called strongly τ-multiplication if every τ-closed submodule is τ-multiplication as an R-module.*

Lemma 3.1.5 *Every multiplication R-module is a τ-multiplication R-module for every torsion theory τ.*

Proof: Let $N \leq M$ such that N is τ-closed. Since M is multiplication, $\exists I \leq R \mid N = M.I$. Taking closures, $N = Cl_\tau^M(N) = Cl_\tau^M(M.I)$. ■

Every ring R is a τ-multiplication R-module as any ideal I can be written as $Cl_\tau^R(R.I)$. For this reason, a ring R is said to be a *τ-multiplication ring* if it is strongly τ-multiplication as an R-module, i.e., if every ideal is τ-multiplication as an R-module.

We have the first examples of τ-multiplication modules in the family of multiplication modules and therefore, projective ideals and cyclic modules (cf. [13]).

Example 3.1.6 An R-module M is called *τ-cocritical* for a hereditary torsion theory τ if M is τ-torsion free and the only τ-closed submodules of M are $\tau(M)$ and M. Then every τ-cocritical module is a τ-multiplication module.

The next example shows that there exist τ-multiplication modules which are not multiplication modules.

Example 3.1.7 Consider the set $\mathbb{Q}$ of rational numbers as a $\mathbb{Z}$-module. Consider the classical torsion theory which is trivially hereditary. If N is a submodule of $\mathbb{Q}$ then its torsion is $\tau(N) = \{x \in N; \exists n \in \mathbb{Z}^* \mid x.n = 0\}$. Then, it is clear that $\mathbb{Q}$ is τ-cocritical and hence τ-multiplication. However it is not multiplication because if $\mathbb{Z} = \mathbb{Q}.I$ for some ideal I of $\mathbb{Z}$, then $\mathbb{Z} = \mathbb{Q}.I = 0$ if $I = 0$, and if not, $\mathbb{Z} = \mathbb{Q}.I = \mathbb{Q}$, a contradiction.

Example 3.1.8 *Every Krull domain is a multiplication ring with respect to the torsion theory corresponding to the canonical topology.*

Let R be a Krull domain. The torsion theory corresponding to the canonical topology is denoted by τ. In order to define it, we will denote by $\mathcal{P}$ the set of prime ideals of height smaller than or equal to 1 and, if I is an ideal of R, $V(I) = \{P \in Spec(R); I \subseteq P\}$. Then, τ is the one that has as Gabriel filter $\mathcal{F} = \{I \leq R; V(I) \cap \mathcal{P} = \emptyset\}$. Recall that if K the field of fractions of the domain R, a *fractional ideal* of R is an R-submodule F of K such that there exists an element $d \in R$ verifying that $Fd \subseteq R$.

Consider A, B, τ-closed ideals of R such that $B \subseteq A$. Since R is τ-noetherian (cf. [105, Corollary XIII.4.6]), by property $C9$, $Cl_\tau^R(C) = \cap_{P \in \mathcal{P}} C_P$ for every C ideal of R.

Let R be an integral domain and let K be its field of fractions. An element $x \in K$ is called *almost integral* over R if there is a fractional ideal A such that $xA \subseteq A$. If the set of all these elements coincide with R, then R is said to be *completely integrally closed.* It is well-known that every Krull domain R is completely integrally closed (cf. [44, Theorem 3.6]) and therefore, $R = (A :_K A)$ for every fractional ideal A ([44, Corollary 3.3]) and moreover, the monoid of fractional ideals is a group ([44, Proposition 3.5]) with the following product: the pair of fractional ideals (A, B) is applied to $(R :_K (R :_K AB))$. Since A is τ-closed (or divisorial), there exists a fractional ideal C of R such that $Cl_\tau^R(C.A) = R$ as for the canonical torsion theory, if A is an integral ideal of R, then $Cl_\tau^R(A) = (R :_K (R :_K A))$. We have

$$B = Cl_\tau^R(B) = Cl_\tau^R(R.B) = Cl_\tau^R(Cl_\tau^R(A.C).B) = Cl_\tau^R(A.B.C)$$

by property K4. Thus there exists an integral ideal $I = CB$ of R such that $B = Cl_\tau^R(IA)$ and hence R is a τ-multiplication ring.

Example 3.1.9 Let τ be a hereditary torsion theory. An R-module M is called *τ-simple* if it is not torsion and $\tau(M)$ (the torsion submodule of M) and M are the only τ-closed submodules of M. Obviously every τ-simple module is a τ-multiplication module as $\tau(M) = Cl_\tau^M(0) = Cl_\tau^M(M.0)$.

Example 3.1.10 Consider $M = \mathbb{Z}_{p^\infty}$ as a $\mathbb{Z}$-module. Let τ be such that $(p) \in \mathcal{F}$. Since M is τ-simple, it is strongly τ-multiplication, but it is not multiplication according to Singh and Mehdi's definition (cf. [77]).

The following example is a ring which is not multiplication but it is τ-multiplication. Moreover, it is not a Krull domain.

Example 3.1.11 Let $R = \Pi_{i\in\mathbb{N}} R_i$ where $R_i = \mathbb{Z}_4$. According to [57, Example 3], R is not a multiplication ring. Set $S = \oplus_{i\in\mathbb{N}} R_i$. Obviously, $S^2 = S$ and it is possible to consider the Gabriel filter $\mathcal{F} = \{A \leq R; S \subseteq A\}$. If B is an ideal of R, then $Cl_\tau^R(B) = (B : S)$ clearly. Denote by e_i the element of R which has the i-th coordinate equal to 1 and the others are 0. If $x \in B$ and B is τ-closed, then each component x_i of x has to verify $x_i.e_i \in B$. Let $x = (x_i)_{i\in\mathbb{N}}$ verifying the preceding condition. If $s \in S$, the product $y.s$ can be seen as a finite sum of elements of B and therefore, it belongs to B. This means that $B = \Pi_{i\in\mathbb{N}}(B \cap R_i)$. Thus τ-closed ideals are ideals of the form $B = \Pi_{i\in\mathbb{N}} B_i$ with $B_i \leq \mathbb{Z}_4$. Let A, B be τ-closed ideals of R such that $A \subseteq B$. Since $\mathbb{Z}_4$ is a multiplication ring (it is uniserial), for every $i \in \mathbb{N}$, there exists an ideal C_i of $\mathbb{Z}_4$ such that $A_i = C_i.B_i$. Consequently, $A = Cl_\tau^R(C.B)$, where $C = \Pi_{i\in\mathbb{N}} C_i$.

Example 3.1.12 If R is a τ-multiplication ring and $\mathcal{F}$ is the corresponding Gabriel filter, then $R \oplus R$ is multiplication with respect to the torsion theory whose Gabriel filter is $\{(I, J); I, J \in \mathcal{F}\}$. Therefore, if D is a Krull domain and τ is the canonical torsion theory, then $D \oplus D$ is a relative multiplication ring and obviously, it is not a Krull domain.

Example 3.1.13 *Every τ-invertible ideal is τ-multiplication.*

An ideal A of a ring R is *τ-invertible* if there exists a fractional ideal B such that $Cl^R_\tau(AB) = R$. Then, if $C \leq R$ verifies $Cl^A_\tau(C) = C$, we have $C = Cl^A_\tau(Cl^R_\tau(AB)C) = Cl^A_\tau(A(BC))$. Moreover, $BC \subseteq BA \subseteq R$. Therefore, A is τ-multiplication as an R-module. In particular, *every ideal belonging to the Gabriel filter is τ-multiplication.*

Definition 3.1.14 *A ring is a τ-SPIR if $Max_\tau(R) = \{P\}$ and every τ-closed ideal of R is a power of the closure of a power of R.*

Example 3.1.15 *Every τ-SPIR is τ-multiplication*

Remark. *i*) There are τ-SPIR's that are not SPIR's. For example, $R = K \oplus K$, K being a field, with Gabriel filter $\{K \oplus 0, K \oplus K\}$.

ii) In a similar way to [69, Lemma 9.8 and Proposition 9.9], we have that if between $A = Cl^R_\tau(A) \leq R$ and $Cl^R_\tau(A^2)$ there is no properly contained τ-closed ideal, then the only τ-closed ideals between $Cl^R_\tau(A)$ and $Cl^R_\tau(A^n)$ are $A, Cl^R_\tau(A^2), ..., Cl^R_\tau(A^n)$. Moreover, if $P \in Max_\tau(R)$ without strict ideals between P and $Cl^R_\tau(P^2)$, then $R/Cl^R_\tau(P^n)$ is a τ-SPIR for the induced torsion theory (which will continue being called τ).

Definition 3.1.16 *Let τ,σ be torsion theories on R with Gabriel filters $\mathcal{F}_\tau$ and $\mathcal{F}_\sigma$ respectively. Then $\tau \wedge \sigma$ is the torsion theory whose Gabriel filter is $\mathcal{F}_\tau \cap \mathcal{F}_\sigma$.*

Proposition 3.1.17 *Let τ and σ be two hereditary torsion theories on R. If $M \in Mod - R$ is τ and σ-multiplication, then M is a $\tau \wedge \sigma$-multiplication module.*

Proof: Let N be $\tau \wedge \sigma$-closed. Then N is τ-closed and σ-closed. By hypothesis and by Lemma 2.2.18, $N = Cl_{\tau}^{M}((N : M).M) = Cl_{\sigma}^{M}((N : M).M)$. So, for every $n \in N$ there exist $I_n \in \mathcal{F}_{\tau}$ and $J_n \in \mathcal{F}_{\sigma}$ such that $I_n.n \subsetneq (N : M).M$ and $J_n.n \subsetneq (N : M).M$. Then $I_n \cap J_n \in \mathcal{F}_{\tau} \cap \mathcal{F}_{\sigma}$ verifying $(I_n \cap J_n).n \subsetneq (N : M).M$. Hence $N = Cl_{\tau\wedge\sigma}^{M}((N : M).M)$.■

Recall that a class $\mathcal{A}$ of R-modules *generates* a torsion theory in the following way:

$$\mathbf{F} = \{F; Hom_R(C, F) = 0\ \forall C \in \mathcal{A}\}$$

$$\mathbf{T} = \{T; Hom_R(T, C) = 0\ \forall T \in \mathcal{A}\}$$

Proposition 3.1.18 *Let M be an R-module. Then, there exists a torsion theory τ such that M is a τ-multiplication module.*

Proof: Consider the family $\mathcal{A} = \{N/(M.(N : M)); N \leq M\}$ and the torsion theory generated by $\mathcal{A}$. Then every element of $\mathcal{A}$ is torsion. So $N = Cl_{\tau}^{N}(M.(N : M))$ for every $N \leq M$. Hence, if N is τ-closed, then $N = Cl_{\tau}^{M}(M.(N : M))$ and therefore M is τ-multiplication.■

Theorem 3.1.19 *Let $M = Cl_{\tau}^{M}(N)$ for some submodule N of M. Let τ a hereditary torsion theory. Then M is τ-multiplication if and only if N is τ-multiplication.*

Proof: Suppose that N is a τ-multiplication R-module. Let A be a τ-closed submodule of M. Then, $A = Cl_{\tau}^{M}(A) \cap M = Cl_{\tau}^{M}(A) \cap Cl_{\tau}^{M}(N) = Cl_{\tau}^{M}(A \cap N) = Cl_{\tau}^{M}(Cl_{\tau}^{N}(A \cap N))$ as $A \cap N \leq N$, which is a τ-multiplication module. Then, there exists an ideal I of R such that $Cl_{\tau}^{N}(A \cap N) = Cl_{\tau}^{N}(N.I)$.

Thus

$$A = Cl_\tau^M(Cl_\tau^N(N.I)) = Cl_\tau^M(Cl_\tau^M(N.I) \cap N) = Cl_\tau^M(N.I)$$

and this module is equal to $Cl_\tau^M(Cl_\tau^M(N).I) = Cl_\tau^M(M.I)$.

Conversely, suppose that M is a τ-multiplication module. Let L be a submodule of N such that $L = Cl_\tau^N(L)$. Since M is τ-multiplication, $Cl_\tau^M(L) = Cl_\tau^M(M.I)$ for some ideal I of R. Thus

$$L = Cl_\tau^M(L) \cap N = Cl_\tau^M(M.I) \cap N = Cl_\tau^M(N.I) = Cl_\tau^N(N.I)$$

and hence N is τ-multiplication.■

Definition 3.1.20 *Let τ be a torsion theory on $Mod-R$. An R-module M is called τ-cyclic if there exists $m \in M$ such that $M = Cl_\tau^M(m.R)$.*

Corollary 3.1.21 *Every τ-cyclic R-module is a τ-multiplication R-module.*

Proof: Straightforward as every cyclic module is multiplication and therefore it is τ-multiplication.■

Corollary 3.1.22 *If M is locally cyclic and τ-finitely generated then M is a τ-multiplication module.*

Proof: Since M is τ-finitely generated as R-module, there exists a submodule F of M such that it is finitely generated and $M = Cl_\tau^M(F)$. As M is locally cyclic (Prüfer), F is a multiplication module and hence a τ-multiplication module. Now, it suffices to apply Theorem 3.1.19.■

We denote by $Max_\tau(M)$ the set of maximal τ-closed submodules of M.

Theorem 3.1.23 *If M is a not τ-torsion τ-multiplication module and τ has enough τ-criticals, then every τ-closed submodule of M is contained in a maximal τ-closed submodule. In this case, $N = Cl_\tau^M(N)$ belongs to $Max_\tau(M)$ if and only if $N = Cl_\tau^M(MP)$ for some $P \in Max_\tau(R)$.*

Proof: Given a τ-closed submodule N of M, the quotient M/N is τ-multiplication. Therefore, it suffices to prove that there exists some maximal τ-closed submodule. Since $M \neq Cl_\tau^M(0)$, we can choose $m \in M - Cl_\tau^M(0)$. It is clear that $(Cl_\tau^M(0) : m) \neq R$ is τ-closed. As τ has enough τ-criticals, there exists $P \in Max_\tau(R)$ such that $(Cl_\tau^M(0) : m) \subseteq P$. Since M is τ-multiplication, $Cl_\tau^M(mR) = Cl_\tau^M(MA)$ for some $A \leq R$. If $M = Cl_\tau^M(MP)$, then $Cl_\tau^M(mR) = Cl_\tau^M(MPA) = Cl_\tau^M(MAP) = Cl_\tau^M(mP)$. It follows that there exists $I \in \mathcal{F}$ such that $mI \subseteq mP$. But $I \not\subseteq P$ implies there is $a \in I \cap (R - P)$ such that $am = 0$. Hence, $a \in (Cl_\tau^M(0) : m)$, a contradiction. Therefore $M \neq Cl_\tau^M(MP)$. If $L = Cl_\tau^M(L)$ verifies $Cl_\tau^M(MP) \subseteq L \subseteq M$ and $L \neq M$, then $(L : M) = Cl_\tau^R(L : M) \neq R$ (otherwise, $L = Cl_\tau^M(M(L : M)) = M$). By hypothesis, there exists $Q \in Max_\tau(R)$ such that $(L : M) \subseteq Q$. Since $MP \subseteq Cl_\tau^M(MP) \subseteq Q$, we have $P = (L : M) = Q$. This yields $L = Cl_\tau^M(M(L : M) = Cl_\tau^M(MP)$ and therefore, $Cl_\tau^M(MP) \in Max_\tau(M)$. ■

In the following proposition we show that being relative multiplication is preserved by homomorphic images. Before that, we need the following Lemma.

Lemma 3.1.24 *Let $f : M \to N$ be a morphism of R-modules. Then, $f(Cl_\tau^M(L)) \subseteq Cl_\tau^{f(M)}(f(L))$ for every submodule L of M.*

Proof: Take an element $y \in f(Cl_\tau^M(L))$. Then, $y = f(x)$ with $x \in Cl_\tau^M(L)$.

Thus there exists an ideal $I \in \mathcal{F}$ such that $xI \subseteq L$. By applying f, we obtain $f(xI) \subseteq f(L)$. Since f is a morphism of R-modules, $yI \subseteq f(L)$ and hence, $y \in Cl_\tau^{f(M)}(f(L))$. ■

Proposition 3.1.25 *If M is a τ-multiplication R-module and $f : M \to N$ is a morphism of R-modules, then $f(M)$ is a τ-multiplication R-module.*

Proof: Let $f(L)$ be a τ-closed submodule of $f(M)$ with $L \leq M$. By hypothesis, $Cl_\tau^M(L) = Cl_\tau^M(M(L : M))$. By applying f, we have $f(Cl_\tau^M(L)) = f(Cl_\tau^M(M(L : M)))$. By Lemma 3.1.24, $f(L) \subseteq f(Cl_\tau^M(L))$
$\subseteq Cl_\tau^{f(M)}(f(L)) = f(L)$. ■

Corollary 3.1.26 *If $M \in R$-Mod is a τ-multiplication module, then M/N is a τ-multiplication R-module for every submodule N of M.*

Proof: Consider the canonical projection from M to M/N. Obviously, it is a morphism of R-modules and moreover, it is surjective. It suffices to apply Theorem 3.1.25 to obtain the wished result. ■

Lemma 3.1.27 *If N is a τ-closed submodule that is not τ-torsion of a τ-multiplication module, then $Hom(M, N) \neq 0$.*

Proof: Since M is τ-multiplication and N is τ-closed, $N = Cl_\tau^M(M(N : M))$. As N is not τ-torsion, there exists $n_0 \in N = Cl_\tau^M(M(N : M)) - Cl_\tau^M(0)$. Therefore, there exists $I \in \mathcal{F}$ such that $n_0 I \subseteq M(N : M)$. For some $a \in I$, we have $n_0 a \neq 0$ because, otherwise $n_0 \in \tau(N)$. Thus $n_0 a = \sum_{i=1}^{i=s} m_i r_i$ with $r_i \in (N : M)$ and $m_i \in M$. Since $n_0 a$ is not zero, there exists i_0 $(1 \leq i_0 \leq s)$ such that $m_{i_0} r_{i_0} \neq 0$. Then, the morphism $\Phi : M \to N$ defined by $\Phi(m) = m r_{i_0}$ takes values, in fact, in N and it is not zero. ■

Definition 3.1.28 *An R-module M is called τ-totally invariant if every τ-closed submodule N is invariant, i. e., $\lambda(N) \subseteq N$ for every $\lambda \in \Lambda = End(M)$.*

Proposition 3.1.29 *Every τ-multiplication module is τ-totally invariant.*

Proof: By Lemma 3.1.24, if $\lambda \in \Lambda$, then $\lambda(Cl_\tau^M(A)) \subseteq Cl_\tau^M(\lambda(A))$ for every $A \leq M$. But $\lambda(yI) = \lambda(y)I = xI \subseteq \lambda(A)$. As a consequence, $x \in Cl_\tau^M(\lambda(A))$.

By hypothesis, we can suppose that $N = Cl_\tau^M(N) = Cl_\tau^M(MJ)$ for some ideal J of R. We have

$$\lambda(N) = \lambda(Cl_\tau^M(MJ)) \subseteq Cl_\tau^M(\lambda(MJ)) = Cl_\tau^M(\lambda(M)J) \subseteq Cl_\tau^M(MJ) = N.$$

■

Proposition 3.1.30 *If $M \in Mod - R$ is not τ-torsion and it is a τ-multiplication module for a hereditary torsion theory τ on R, then M is not isomorphic to $X \oplus X$ for any submodule X of M.*

Proof: We suppose that M is isomorphic to $X \oplus X$. The submodule $Cl_\tau^M(X \oplus 0)$ is a τ-closed submodule of M. By the hypothesis we have $Cl_\tau^M(X \oplus 0) = Cl_\tau^M((X \oplus X).I)$ for some ideal I of R. Then

$$(0 \oplus X).I \subseteq (X \oplus X).I \subseteq Cl_\tau^M(X \oplus 0) = Cl_\tau^M((X \oplus X).I)$$

So , for every $x \in X$ and for every $i \in I$ there exists an ideal $J \in \mathcal{F}$ such that $(0 \oplus x.i).J \subseteq X \oplus 0$. Hence $x.i \in Cl_\tau^X(0)$ and therefore $X.I \subseteq Cl_\tau^X(0)$. Then,

$$X \oplus 0 \subseteq Cl_\tau^M((X \oplus X).I) \subseteq Cl_\tau^M(X.I \oplus X.I) \subseteq Cl_\tau^M(Cl_\tau^X(0) \oplus Cl_\tau^X(0))$$

Therefore $X = Cl_\tau^X(0)$ and X is τ-torsion. Since the τ-torsion class is closed under extensions, M is τ-torsion, a contradiction.■

Theorem 3.1.31 *$M \in Mod - R$ is a τ-multiplication module if and only if for every element $m \in M$ there exists an ideal I of R such that $Cl_\tau^M(m.R) = Cl_\tau^M(M.I)$.*

Proof: The necessity is clear. Conversely, suppose that M has the stated property and N is a τ-closed submodule of M. For every $n \in N$ there exists an ideal I_n such that $Cl_\tau^M(n.R) = Cl_\tau^M(M.I_n)$.

Consider the ideal $I = \sum_{n \in N} I_n$. Then, by the property $K.3$, we have

$$N = \sum_{n \in N} Cl_\tau^M(n.R) = \sum_{n \in N} Cl_\tau^M(M.I_n) \subseteq \sum_{n \in N} Cl_\tau^M(M.I) = Cl_\tau^M(M.I)$$

Hence $N \subseteq Cl_\tau^M(M.I)$.

On the other hand $M.I \subseteq \sum_{n \in N}(M.I_n)$. So, by properties $K.3$ and PK2, we have

$$Cl_\tau^M(M.I) \subseteq Cl_\tau^M(\sum_{n \in N}(M.I_n)) = Cl_\tau^M(\sum_{n \in N} Cl_\tau^M(M.I_n))$$

and this module is equal to $Cl_\tau^M(\sum_{n \in N} Cl_\tau^M(n.R)) = Cl_\tau^M(N) = N$. Hence $Cl_\tau^M(M.I) \subseteq N$ and therefore we have equality. It follows that M is τ-multiplication. ■

Now, we give some chain condition properties.

Proposition 3.1.32 *If R is τ-noetherian (τ-artinian) and M is a τ-multiplication module, then M is τ-noetherian (τ-artinian).*

Proof: Show that M has A.C.C. on τ-closed submodules. In fact, we consider $N_1 \leq N_2 \leq ...$ with $N_i \leq M$ τ-closed ($i \in I$). Then $N_i = Cl_\tau^M((N_i : M).M)$ by Lemma 2.2.18. But $Cl_\tau^M(N_i : M) = (Cl_\tau^M(N_i) : M)$ from Proposition 1.1.6 and therefore $(N_i : M)$ is a τ-closed ideal for every $i \in I$. Moreover, $(N_1 : M) \leq (N_2 : M) \leq ...$ By hypothesis, there exists i such that $(N_i : M) = (N_{i+1} : M) = (N_{i+2} : M) = ...$ and hence

$$N_i = Cl_\tau^M((N_i : M).M) = Cl_\tau^M((N_{i+1} : M).M) = N_{i+1} = ...$$

and therefore M is τ-noetherian. For the artinian case the proof is analogous.■

The converse result is false. In fact, consider the ring $\mathbb{Z}$ which is τ-noetherian for any τ. Let $M = \mathbb{Z} \oplus \mathbb{Z}$. M is τ-noetherian but it is not τ-multiplication for any torsion theory τ different from the trivial one by Proposition 3.1.30.

Example 3.1.33 Let $M = \mathbb{Z}[x_1, x_2, \cdots]$ the $\mathbb{Z}$-module consisting of all polynomials in infinite indeterminates $x_1, x_2, \cdots$ By [105, Corollary VI.6.15], every torsion theory on $\mathbb{Z}$ is semicentered. If τ is different from the trivial one, then M is not τ-noetherian, obviously. By applying Proposition 3.1.32, M is not τ-multiplication.

3.2 Localization with respect to torsion theories

We study localization with respect to τ-closed prime ideals and with respect to a torsion theory. We mainly work with semicentered torsion theories or with perfect torsion theories. In this case, the concepts of ring and module of quotients with respect to a torsion theory will be managed. We shall characterize τ-multiplication τ-finitely generated modules, when τ is semicentered in several different ways.

The next result characterizes, for perfect torsion theories, multiplication modules relative to τ in terms of the corresponding module and ring of quotients with respect to τ.

Proposition 3.2.1 *Let R be a commutative ring. Let τ a perfect torsion theory on $Mod - R$. Then the R_τ-module M_τ is multiplication if and only if the R-module M is τ-multiplication.*

Proof: We assume that M_τ is multiplication as an R_τ-module. Let N be a τ-closed submodule of M. By the bijective correspondences of [105, 9.4.3] and [105, 11.3.7] and the isomorphism of [105, 11.3.4 (d)], N corresponds to N_τ and $N \otimes_R R_\tau$. By the hypothesis, there exists $i(P).R_\tau \leq R_\tau$ with I such that $i(I).R_\tau = i(I)$ verifying $N_\tau = M_\tau.i(I).R_\tau$. Then $Cl_\tau^M(M.I)$ is a τ-closed submodule of M and it is applied in $Cl_\tau^M(M.I) \otimes_R R_\tau = M.I \otimes_R R_\tau$. In fact, if $y \in Cl_\tau^M(M.I) \otimes R_\tau$ then y is a finite sum of elements of the form $x \otimes \bar{r}$ where $\bar{r} \in R_\tau$ and $x \in Cl_\tau^M(M.I)$. So there exists $J \leq R$ in the Gabriel filter such that $x.J \subseteq M.I$.

We have $(x \otimes \bar{r}).R_\tau = (x \otimes \bar{r}).i(J).R_\tau = x.J \otimes \bar{r}.R_\tau \subseteq M.I \otimes_R R_\tau$. Therefore $N = Cl_\tau^M(M.I)$ and M is a τ-multiplication module.

Conversely, suppose M is a τ-multiplication module. Let L be an R_τ-submodule of M_τ. Then L is associated with a τ-closed submodule of M, N, because of the mentioned correspondences.

By the hypothesis, $N = Cl_\tau^M(M.A)$ for some ideal A of R. This submodule corresponds to $Cl_\tau^M(M.A) \otimes_R R_\tau = M.A \otimes_R R_\tau$ (by the above argument). However, $M.A \otimes_R R_\tau = (M \otimes_\tau R_\tau).i(A).R_\tau$ and this module corresponds to $M_\tau.i(A).R_\tau$. Therefore $L = M_\tau.i(A).R_\tau$.■

Definition 3.2.2 *An R-module M is said to be τ-projective if M_τ is projective as an R_τ-module.*

Corollary 3.2.3 *Let τ be a perfect torsion theory. If I is a τ-projective ideal of R, then I is τ-multiplication as an R-module.*

Proof: By hypothesis, I_τ is a projective ideal and therefore it is multiplication (cf. [104, Theorem 1]). By Proposition 3.2.1, I is a τ-multiplication ideal.■

The result does not hold for τ-projective modules. If M is a projective module and every τ-projective module is τ-multiplication, then $M \oplus M$ is a projective module which is not τ-multiplication by Proposition 3.1.30, a contradiction.

Remark. Notice that the ring of quotient of Example 3.1.11 is

$$R_\tau = Hom_R(S, R) = \prod_{i \in \mathbb{N}} Hom_R(R_i, R) = \prod_{i \in \mathbb{N}} \prod_{j \in \mathbb{N}} Hom_R(R_i, R_j),$$

which is isomorphic to $\prod_{i \in \mathbb{N}} Hom_R(R_i, R_i) = R$. Since R_τ is not a multiplication ring but R is τ-multiplication, it is proved that Proposition 3.2.1 is not always true if τ is not perfect.

Let τ be a hereditary torsion theory. For every prime ideal P, consider $S = R - P$ and $\mathcal{F}_S = \{I \leq R; I \cap S \neq \emptyset\}$. Let M be an R-module. We denote by Cl_S^M the closure corresponding to $\mathcal{F}_S$ and by R_S and M_S the corresponding rings and modules of quotients.

Lemma 3.2.4 *For every prime ideal $P \notin \mathcal{F}_\tau$ we consider $S = R - P$. Then $Cl_\tau^M(Cl_S^M(A)) = Cl_S^M(Cl_\tau^M(A)) = Cl_S^M(A)$ for each submodule A of M.*

Proof: If $I \in \mathcal{F}_\tau$, the Gabriel filter for τ, then $I \not\subseteq P$ for all $P \notin \mathcal{F}_\tau$. So $I \in \mathcal{F}_S$ for all $S = R - P$. Hence $Cl_\tau^M(A) \subseteq Cl_S^M(A)$. If in this formula we consider $Cl_S^M(A)$ instead of A then $Cl_\tau^M(Cl_S^M(A)) \subseteq Cl_S^M(A)$. Obviously,

$Cl_S^M(A) \subseteq Cl_\tau^M(Cl_S^M(A))$ as $A \subseteq Cl_\tau^M(A)$. By applying Cl_S^M in the first formula, we obtain $Cl_S^M(Cl_\tau^M(A)) \subseteq Cl_S^M(A)$.

Take $x \in Cl_S^M(A)$. Then there exists $I \in \mathcal{F}_S$ such that $x.I \subseteq Cl_\tau^M(A)$. So there exists $c \in R - P$ with $x.c \in Cl_\tau^M(A)$. Thus there exists $J \in \mathcal{F}_\tau$ with $x.c.J \subseteq A$. Moreover $c.J \in \mathcal{F}_S$ because $(R - P) \cap J$ is non-empty. Thus $Cl_S^M(A) \subseteq Cl_S^M(Cl_\tau^M(A))$. Hence we have the equalities.■

It is known that for finitely generated modules, the property of being a multiplication module is local. The following theorem is a generalization (cf. [13, Lemma 2]).

Theorem 3.2.5 *Let τ be a semicentered torsion theory on R.*

1. *If M is a τ-multiplication module then the R_P-module M_P is multiplication for all $P \in K(\tau)$.*

2. *If M is finitely generated then M is τ-multiplication if and only if M_P is multiplication for all $P \in K(\tau)$.*

Proof: Suppose M that M is a τ-multiplication module. Let $P \in K(\tau)$. Let $S = R - P$. Let N be a S-closed submodule of M. Then $Cl_\tau^M(N) =$

$Cl_\tau^M(Cl_S^M(N)) = Cl_\tau^M(N) = N$ by Lemma 3.2.4. By hypothesis, there exists an ideal A of R such that

$$N = Cl_S^M(N) = Cl_\tau^M(M.A) = Cl_\tau^M(Cl_S^M(M.A)) = Cl_S^M(M.A).$$

by Lemma 3.2.4. Thus M is a multiplication module relative to the torsion theory of the filter $\mathcal{F}_S$. However this torsion theory, τ_S is perfect. By Proposition 3.2.1, M_S is multiplication as an R_τ-module. But $M_\tau = M_P$ and $R_S = R_P$.

Now, we suppose that M is finitely generated and that for all $P \in K(\tau)$, M_P is multiplication as an R_P-module. By Proposition 3.2.1, M is a τ_S-multiplication module (with $S = R - P$). Let N be a τ-closed submodule of M. Consider $y \in N$ and $K_y = \{r \in R; y.r \in Cl_\tau^M(M.(N : M))\}$. If for all y in N, K_y belongs to the filter then $N = Cl_\tau^M(M.(N : M))$. Suppose there exists $y \in N$ such that $K_y \notin \mathcal{F}$. Then there exists $P \in K(\tau)$ such that $K_y \subseteq P$. Since M_P is multiplication as R_P-module we have $N \subseteq Cl_S^M(N) = Cl_S^M(M.(Cl_S^M(N) : M)) = Cl_S^M(M.Cl_S^M(N : M))$ by Property PC3. For $y \in N$, there exists an ideal $I \in \mathcal{F}_S$ such that $y.I \subseteq M.Cl_S^M(N : M)$. Therefore, there exists $c \in (R - P) \cap I$ such that $y.c = m_1.a_1 + m_2.a_2 + ... + m_n.a_n$ with $m_l \in M$ and $a_l \in Cl_S^M(N : M)(1 \leq l \leq n)$. Hence, there exist $J_l \in \mathcal{F}_S$ such that $a_l.J_l \subseteq (N : M)$. Thus, there exists $J = J_1.J_2...J_n \in \mathcal{F}_S$ such that $y.c.J \in M.(N : M)$. Then, there exists $d \in J \cap (R - P)$ such that $y.c.d \in M.(N : M) \subseteq Cl_\tau^M(M.(N : M))$. However, $c.d \in (R - P) \cap K_y$, a contradiction. Therefore, M is a τ-multiplication R-module.■

Corollary 3.2.6 *If τ is semicentered and M is τ-finitely generated, M is τ-multiplication if and only if M_P is cyclic as an R_P-module for every $P \in$*

$K(\tau)$.

Proof: By Theorem 3.2.5, M is τ-multiplication if and only if M_P is multiplication as an R_P-module. Since M is τ-finitely generated, M_P is finitely generated. By [13, Proposition 6], M_P is locally cyclic and in particular, M_P, as by localizing in the maximal ideal of R_P, we obtain M_P. ■

The following characterization is immediate from Theorem 3.2.5.

Proposition 3.2.7 *If R is τ-noetherian and τ is semicentered, then R is a τ-multiplication ring if and only if R_P is a multiplication ring for each $P \in K(\tau)$.*

Recall that a *special primary ring* is a ring with a unique maximal ideal verifying every proper ideal is a power of the maximal ideal. [69, Proposition 9.9] establishes that if M is a maximal ideal of a ring R such that there is no ideal of R strictly between M and M^2, then for every positive integer n, R/M^n is a special primary ring (or more briefly, SPIR). It follows that the ring $\mathbb{Z}_{p^n}$ is a SPIR for every prime p and every positive integer n. Another kind of example are rings of the form $K[X]/(X^n)$ where K is a field.

Corollary 3.2.8 *Let τ be a semicentered torsion theory. Then, τ-multiplication τ-noetherian rings are those whose localized ones in the τ-closed prime ideals are discrete valuation domains or special primary rings.*

Proof: In the τ-noetherian case, by Theorem 3.2.5, every ideal I of R is τ-multiplication if and only if I_P is multiplication as an R_P-module for every $P \in K(\tau)$, i.e., if and only if every R_P is a multiplication ring. But every R_P

is noetherian. By [51, Theorem 39.2], it is a finite direct sum of Dedekind domains and special primary rings. Since R_P is local and noetherian, Dedekind domains must be discrete valuation domains. The corollary follows. ■

Recall that an integral domain is an *almost Dedekind domain* if R_P is a Dedekind domain for every prime ideal P.

Corollary 3.2.9 *Every τ-noetherian almost Dedekind domain, τ being semicentered, is a τ-multiplication ring.*

Proof: Let R be an almost Dedekind domain. Then, if $P \in Spec_\tau(R)$, R_P is a Dedekind domain and therefore, it is a multiplication ring. Since it is local and noetherian, it must be a discrete valuation domain. By Corollary 3.2.8, R is a τ-multiplication ring. ■

The following definition can be found in [114, p. 37].

Definition 3.2.10 *An R-module M is called τ-quasiprojective if M_P is a projective R_P-module for each $P \in K(\tau)$.*

Corollary 3.2.11 *Let τ be a semicentered torsion theory on R. Every finitely generated τ-quasiprojective ideal M is a τ-multiplication R-module.*

Proof: Let $P \in K(\tau)$. Then M_P is projective as an R_P-module. Therefore, M_P is a multiplication module (cf. [104, Theorem 1]). By Theorem 3.2.5, M is τ-multiplication.■

Remark. If R has sufficent τ-critical ideals, then this theorem and the next one hold if we replace $K(\tau)$ by $Max_{\mathcal{F}}(R)$.

Let M be an R-module. If $P \in Spec(R)$ then we denote by $T_P(M)$ the set $\{m \in M; \exists c \in R - P \mid m.c = 0\}$.

Definition 3.2.12 *An R-module M is called P-torsion if* $M = T_P(M)$.

Theorem 3.2.13 *If τ is semicentered on R and M is an R-module, then M is τ-multiplication if and only if M is P-torsion or for some $c \in R - P$ and $m_0 \in M$, $M.c \subseteq Cl_\tau^M(m_0.R)$ for all $P \in K(\tau)$.*

Proof: Suppose that M is P-torsion or for some $c \in R - P$ and $m_0 \in M$, $M.c \subseteq Cl_\tau^M(m_0.R)$ for all $P \in K(\tau)$ and let N be a τ-closed submodule of M. Let $y \in N$. For every y we consider

$$K_y = \{r \in R;\ y.r \in Cl_\tau^M(M.(N : M))\}.$$

If $K_y \in \mathcal{F}$, the Gabriel filter, for every $y \in N$, then $y.K_y \subseteq Cl_\tau^M(M.(N : M))$ and hence $y \in Cl_\tau^M(M.(N : M))$ and therefore $N = Cl_\tau^M(M.(N : M))$.

Suppose there exists $y \in N$ such that $K_y \notin \mathcal{F}$. Then there exists $P \in K(\tau)$ such that $K_y \subseteq P$ since τ is semicentered. If $M = T_P(M)$ then $y.c = 0$ for some $c \in R - P$ and so $c \in K_y \subseteq P$, a contradiction. Suppose $M.c \subseteq Cl_\tau^M(m_0.R)$ for some $c \in R - P$ and $m_0 \in M$. Then $Cl_\tau^M(N.c)$ is τ-closed in $Cl_\tau^M(m_0.R)$ which is τ-multiplication because of Corollary 3.1.21. Thus $Cl_\tau^M(N.c) = Cl_\tau^M(m_0.J)$. By Proposition 1.1.6,

$$J = (Cl_\tau^M(N.c) : Cl_\tau^M(m_0.R)) = (Cl_\tau^M(N.c) : m_0.R).$$

Then,

$$\begin{aligned} M.J.c \subseteq Cl_\tau^M(M.J.c) &= Cl_\tau^M(M.c.J) \subseteq Cl_\tau^M(Cl_\tau^M(m_0.R).J) \\ &= Cl_\tau^M(m_0.J.R) = Cl_\tau^M(m_0.J) = Cl_\tau^M(N.c) \subseteq N. \end{aligned}$$

Thus $J.c \subseteq (N : M)$. However,

$$y.c^2 \in Cl_\tau^M(N.c^2) = Cl_\tau^M(m.J.c) \subseteq Cl_\tau^M(M.(N : M))$$

Therefore $c^2 \in K_y \subseteq P$, a contradiction. Thus $N = Cl_\tau^M(M.(N : M))$ and M is τ-multiplication by Lemma 2.2.18.

Now, we suppose M is τ-multiplication and $P \in K(\tau)$. Suppose $M = Cl_\tau^M(M.P)$. Then for every $m \in M$, $Cl_\tau^M(M.I) = Cl_\tau^M(m.R)$ for some ideal I of R. Thus

$$m \in Cl_\tau^M(Cl_\tau^M(M.P).I) = Cl_\tau^M(M.P.I) = Cl_\tau^M(Cl_\tau^M(M.I).P) = Cl_\tau^M(m.P).$$

Therefore there exists $I \in \mathcal{F}$ such that $m.I \subseteq m.P$. We can choose $i \in I \cap (R - P)$ because $P \notin \mathcal{F}$ and have $m.i = m.p$ and hence $m.(p - i) = 0$. Clearly, $p - i \notin P$. Thus there exists $c = p - i \in R - P$ such that $m.c = 0$. Hence $M = T_P(M)$.

If $M \neq Cl_\tau^M(M.P)$, then there exists $m_0 \in M - Cl_\tau^M(M.P)$. So, $m_0 \in Cl_\tau^M(m_0.R) = Cl_\tau^M(M.B)$ for some ideal B of R. If $B \subseteq P$, then $m_0 \in Cl_\tau^M(M.B) \subseteq Cl_\tau^M(M.P)$, a contradiction. There exists $b \in (R - P) \cap B$ such that $M.b \subseteq M.B \subseteq Cl_\tau^M(M.B) = Cl_\tau^M(m_0.R)$. ■

Corollary 3.2.14 *If τ is semicentered on a commutative ring R, then M is a τ-multiplication module if and only if for every $P \in K(\tau)$, $M = T_P(M)$ or there exists a τ-multiplication submodule N of M and $c \in R - P$ such that $M.c \subseteq N$.*

Proof: The necessity is immediate because of Theorem 3.2.13.

Conversely, suppose that $P \in K(\tau)$. If $M = T_P(M)$, then by the theorem, M is τ-multiplication. Suppose $M.c \subseteq N$ for some submodule N of M and

$c \in R - P$. Hence, if $T_P(M) \neq M$, then $N \neq T_P(N)$. Applying the theorem, there exists $d \in R - P$ and $n \in N$ such that $N.d \subseteq Cl_\tau^N(n.R) \subseteq Cl_\tau^M(n.R)$ as N is a τ-multiplication module. Then, we have $M.c.d \subseteq N.d \subseteq Cl_\tau^M(n.R)$ and hence M is τ-multiplication.■

3.3 Construction of relative multiplication modules

The aim of this section is to study operations such as intersection, sum, direct sum, multiplication, etc, between multiplication modules relative to a torsion theory as a way to obtain new τ-multiplication modules. The achieved results will show that relative multiplication modules constitute an important class of modules. As an application, all τ-multiplication modules over a Dedekind domain will be found.

We start studying sum and intersection. The following result generalizes [102, Lemma 7].

Proposition 3.3.1 *If M is a τ-multiplication R-module and $M = \sum_{i \in I} M_i$, then $N = Cl_\tau^M(\sum_{i \in I}(N \cap M_i))$ for each N τ-closed submodule of M.*

Proof: Since M is τ-multiplication module we have

$$N = Cl_\tau^M(M(N : M)) = Cl_\tau^M((\sum_{i \in I} M_i)(N : M))$$
$$\subseteq Cl_\tau^M((\sum_{i \in I} M_i)(N : M)).$$

Thus we can write N as $Cl_{\tau}^{M}(\sum_{i\in I}(M_i \cap N))$. Therefore, we have $N = Cl_{\tau}^{M}(\sum_{i\in I}(N \cap M_i))$. ■

Proposition 3.3.2 *Let τ be a semicentered torsion theory on R and let $M = \sum_{i\in I} Cl_{\tau}^{M}(m_i R)$ for some elements $m_i \in M$ $(i \in I)$. M is a τ-multiplication module if and only if there exists an ideal J_i $(i \in I)$ such that $Cl_{\tau}^{M}(m_i R) = Cl_{\tau}^{M}(MJ_i)$ for each $i \in I$.*

Proof: The necessity is clear.

Conversely, suppose the existence of such ideals J_i and let $P \notin \mathcal{F}$. If there exists $i \in I$ such that $J_i \not\subseteq P$, then, by hypothesis, $MJ_i \subseteq Cl_{\tau}^{M}(m_i R)$. Hence, there exists $c \in J_i - P \subseteq R - P$ such that $Mc \subseteq Cl_{\tau}^{M}(m_i R)$. If not, we have $J_i \subseteq P$ for all $i \in I$. Thus $Cl_{\tau}^{M}(m_i R) = Cl_{\tau}^{M}(MJ_i) \subseteq Cl_{\tau}^{M}(MP)$ for all $m_i \in M$ and $M = Cl_{\tau}^{M}(MP)$ by hypothesis. Therefore, there exists an ideal J_i of R such that $Cl_{\tau}^{M}(m_i R) = Cl_{\tau}^{M}(MJ_i) = Cl_{\tau}^{M}(.Cl_{\tau}^{M}(MP)J_i) = Cl_{\tau}^{M}(MPJ_i) = Cl_{\tau}^{M}(Cl_{\tau}^{M}(MJ_i)P) = Cl_{\tau}^{M}(m_i P)$. Thus $m_i \in Cl_{\tau}^{M}(P.m_i)$. Therefore, there exists $H_i \in \mathcal{F}$ such that $m_i H_i \subseteq m_i P$ and moreover, $H_i \not\subseteq P$. Hence, there exists $h - p \in R - P$ such that $m_i(h-p) = 0$ and hence $m_i \in T_P(M)$. Then $m_i R \subseteq T_P(M)$. Obviously, $Cl_{\tau}^{M}(T_P(M)) = T_P(M)$. Thus $Cl_{\tau}^{M}(m_i R) \subseteq T_P(M)$ and therefore $M = T_P(M)$. By Theorem 3.2.13, M is τ-multiplication.■

It is straightforward from Proposition 3.3.2 that every τ-cyclic module is a τ-multiplication module.

Example 3.3.3 Consider the $\mathbb{Z}$-module $M = \mathbb{Z}_{p^\infty} = \{\frac{1}{p^i} + \mathbb{Z}; i \in \mathbb{Z}^*\} \subset \mathbb{Q}/\mathbb{Z}$ where p is a prime. If $(p) \in \mathcal{F}$, then $M = Cl_{\tau}^{M}((\frac{1}{p} + \mathbb{Z})\mathbb{Z})$ and by Proposition 3.3.2, M is a τ-multiplication module.

If $(p) \notin \mathcal{F}$, then M is not τ-noetherian. By Proposition 3.1.32, it is not τ-multiplication.

Proposition 3.3.4 *Let τ be a semicentered torsion theory on R. If I is a τ-multiplication ideal of R and M is a τ-multiplication R-module, then MI is a τ-multiplication R-module.*

Proof: Consider $P \in K(\tau)$. It is clear that if $I = T_P(I)$ or $M = T_P(M)$, then $T_P(MI) = MI$. If $I \neq T_P(I)$ and $M \neq T_P(M)$, then there exist $c, d \in R - P$ such that $Ic \subseteq Cl_\tau^R(aR)$ and $Md \subseteq Cl_\tau^M(mR)$ for some $a \in R$ and $m \in M$. Therefore $MIcd \subseteq Cl_\tau^R(aR).Cl_\tau^M(mR) \subseteq Cl_\tau^M(maR)$. By Theorem 3.2.13, MI is a τ-multiplication module.■

The next result answers the question of when the sum of τ-multiplication modules is τ-multiplication. It is the analogous one to [102, Theorem 2].

Theorem 3.3.5 *Let τ be a semicentered torsion theory on R. Let M_i $(i \in I)$ be a family of τ-multiplication τ-closed submodules of an R-module M such that $M = \sum_{i\in I} M_i$. Let $A = \sum_{i\in I}(M_i : M)$. Then the following conditions are equivalent:*

1. *M is a τ-multiplication module.*

2. *$M_i = Cl_\tau^M(M(M_i : M))$ for all $i \in I$.*

3. *$ann(m) + A \in \mathcal{F}$ for all $m \in M$.*

4. *For every $P \in K(\tau)$ either $M = T_P(M)$ or there exist $z \in \cup_{i\in I} M_i$ and $c \in R - P$ such that $Mc \subseteq Cl_\tau^M(zR)$.*

Proof: $1 \Rightarrow 2$ is clear. Now suppose 2 holds. Suppose that $m \in M$ and $ann(m) + A \notin \mathcal{F}$. Since τ is semicentered, there exists $P \in K(\tau)$ such that $ann(m) + A \subseteq P$. So, $(M_i : M) \subseteq P$ for all $i \in I$. Hence $M(M_i : M) \subseteq MP$ and we have $M_i = Cl_\tau^M(M(M_i : M)) \subseteq Cl_\tau^M(MP)$. Thus $M = Cl_\tau^M(MP)$. As $m \in M$, then $m = x_1 + x_2 + ... + x_n$ with $x_i \in M_i$ for $i \in \{1, 2, .., n\}$. Since M_i is a τ-multiplication module, we have $Cl_\tau^M(x_i R) = Cl_\tau^M(M_i.B_i)$ for some ideal B_i of R. Then, by the same argument as in the proof of Proposition 3.3.2, $Cl_\tau^M(x_i R) = Cl_\tau^M(x_i P)$. Therefore there exists $K_i \in \mathcal{F}$ such that $x_i.K_i \subseteq x_i P$ for each $i \in \{1, 2, ..., n\}$. Hence there exists $c_i \in R - P$ such that $x_i.c_i = 0$ for each $i \in \{1, 2, ..., n\}$. Therefore there exists $c \in R - P$ such that $mc = 0$. But then, $c \in ann(m)$ which contradicts $ann(m) \subseteq P$. Thus 3 is satisfied.

$3 \Rightarrow 4$. Let $P \in K(\tau)$ and suppose that $T_P(M) \neq M$. Then, there exists $m \in M$ such that $ann(m) \subseteq P$. By condition 3, $A \nsubseteq P$. Therefore, there exists $i \in I$ such that $(M_i : M) \nsubseteq P$. Hence there exists $c \in R - P$ such that $Mc \subseteq M_i$. Moreover, $M_i \neq T_P(M_i)$ because if not, then $Mc \subseteq T_P(M_i)$ and M would be P-torsion. By Theorem 3.2.13, there exist $c' \in R - P$ and $y \in M_i$ such that $M_i c' \subseteq Cl_\tau^M(yR)$. Therefore $Mc.c' \subseteq M_i c' \subseteq Cl_\tau^M(yR)$ and $c.c' \in R - P$.

$4 \Rightarrow 1$ by Theorem 3.2.13.■

Remark 3.3.6 The result is still true if $M = Cl_\tau^M(\sum_{i \in I} M_i)$.

Corollary 3.3.7 *Let τ be a semicentered torsion theory on R. Let M_i $(i \in I)$ be a family of τ-multiplication τ-closed submodules of an R-module M. If $\sum_{i \in I}(M_i : M) \in \mathcal{F}$, then M is a τ-multiplication module.*

Proof: We have

$$\begin{aligned} M &= Cl_\tau^M(M(\sum_{i\in I}(M_i : M))) \\ &\subseteq Cl_\tau^M(M(\sum_{i\in I}(M_i : M))) \\ &\subseteq Cl_\tau^M(\sum_{i\in I} M_i). \end{aligned}$$

Since $ann(m) + \sum_{i\in I}(M_i : M) \in \mathcal{F}$, it suffices to apply Theorem 3.3.5 and Remark 3.3.6.■

Under these conditions we denote by $A = \sum_{i\in I}(M_i : M)$.

Corollary 3.3.8 *Let τ be a semicentered torsion theory on R. Let M_i ($i \in I$) be a family of τ-closed τ-multiplication finitely generated submodules of M. If $M = \sum_{i\in I} M_i$, then M is a τ-multiplication module if and only if $ann(M_i) + A \in \mathcal{F}$.*

Proof: Suppose that M is τ-multiplication and $M_i =< x_1, ..., x_n >$ (n depending on i). From Theorem 3.3.16, $ann(x_j) + A \in \mathcal{F}$ $(1 \leq j \leq n)$. Hence

$$[ann(x_1) \cap ... \cap ann(x_n)] + A \supseteq \Pi_{j=1}^n (ann(x_j) + A) \in \mathcal{F}.$$

Therefore $ann(M_i) + A = [\cap_{j=1}^n ann(x_j)] + A \in \mathcal{F}$ for all $i \in I$.

Now, suppose that $ann(M_i) + A \in \mathcal{F}$ for all $i \in I$. Let $m \in M$. Since $M = \sum_{i\in I} M_i$, $m = m_1 + ... + m_r$ with $m_j \in M_j$ $(1 \leq j \leq r)$. Since $ann(m_j) \supseteq ann(M_j)$, $ann(m_i) + A \in \mathcal{F}$ for $1 \leq j \leq n$. Moreover, $ann(m) + A = [\cap_{j=1}^n ann(m_j)] + A \supseteq \Pi_{j=1}^n (ann(m_j) + A) \in \mathcal{F}$. Therefore $ann(m) + A \in \mathcal{F}$ for all $m \in M$. By Theorem 3.3.5, M is a τ-multiplication module.■

Example 3.3.9 Let $M = \oplus_{n=1}^{\infty} C_{p^n}$ where C_{p^n} is the cyclic group of order p^n and p a prime integer. M is a $\mathbb{Z}$-module. Every C_{p^i} is cyclic and therefore it is τ-multiplication.

If $(p) \notin \mathcal{F}$, then every C_{p^i} is τ-closed. Moreover, $(C_{p^i} : M) = 0$. Thus $A = 0$. It holds $ann(C_{p^i}) + A = (p^i) \notin \mathcal{F}$ for all $i \geq 0$. By Corollary 3.3.8, M is not τ-multiplication.

Suppose that $(p) \in \mathcal{F}$. Let $x = x_{i_1} \oplus \cdots \oplus x_{i_n} \in M$ where each $x_{i_j} \in C_{p^j}$. We have $(p^{i_n}).x = 0$. Hence M is τ-torsion. Therefore M is τ-multiplication in this case.

Corollary 3.3.10 *Let τ be a semicentered torsion theory on R. Let $M = \sum_{i \in I} M_i$, M_i being a τ-closed τ-multiplication finitely generated submodule of M for all $i \in I$. M is τ-finitely generated if and only if there exists a finite subset $J \subseteq I$ such that $\sum_{i \in J}(M_i : M) \in \mathcal{F}$.*

Proof: Since M is τ-finitely generated, there exists a finitely generated submodule F of M such that $M = Cl_{\tau}^{M}(F)$. Therefore there exists a finite subset J of I such that $M = Cl_{\tau}^{M}(\sum_{i \in J} M_i)$. By Theorem 3.3.5 and Remark 3.3.6, $ann(m) + \sum_{i \in J}(M_i : M) \in \mathcal{F}$ for all $m \in M$, in particular, for all $m \in F$. As F is finitely generated, it holds that $ann(F) + \sum_{i \in J}(M_i : M) \in \mathcal{F}$. However, $ann(F) \subseteq (M_i : M)$ for all $i \in J$. Hence $\sum_{i \in J}(M_i : M) \in \mathcal{F}$.

Conversely, suppose that $\sum_{i \in J}(M_i : M) \in \mathcal{F}$ for some finite subset J of I. By Theorem 3.3.5, M is τ-multiplication. Moreover,

$$M = Cl_{\tau}^{M}(M(\sum_{i \in J}(M_i : M))) = Cl_{\tau}^{M}(\sum_{i \in J} M(M_i : M)) = Cl_{\tau}^{M}(\sum_{i \in J} M_i).$$

Therefore M is τ-finitely generated.■

Corollary 3.3.11 *Let $K, L_1, ..., L_n$ be τ-closed submodules of M, τ being semicentered. If $K, K + L_i$ $(1 \leq i \leq n)$, $L_1 \cap ... \cap L_n$ are τ-multiplication modules, then $K + (L_1 \cap ... \cap L_n)$ is a τ-multiplication module.*

Proof: Let $P \in K(\tau)$. Call $L = L_1 \cap ... \cap L_n$. Clearly, L is τ-closed. Suppose that $T_P(K + L) \neq K + L$. Then $T_P(K + L_i) \neq K + L_i$ for each $1 \leq i \leq n$. Consider $A = (K : (K + L_i)) + (L_i : (K + L_i))$. By applying Theorem 3.3.5 to $K + L_i$, we obtain $A \not\subseteq P$ as there exists $m \in K + L_i$ such that $ann(m) + A \subseteq P$. However, $A = (K : L_i) + (L_i : K)$. Since $(K : L_i) \subseteq (K : L)$, we deduce that $(K : L) + (L_i : K) \not\subseteq P$ for $1 \leq i \leq n$. Hence $(K : L) + (L : K) = (K : L) + [(L_1 : K) \cap ... \cap (L_n : K)] \not\subseteq P$. Therefore there exists $c' \in R - P$ such that $c' = a_1 + a_2$ with $a_1 \in (K : L)$ and $a_2 \in (L : K)$. Thus there exists $c \in R - P$ (a_1 or a_2) such that $c \in (K : L)$ or $c \in (L : K)$. Hence $Lc \subseteq K$ or $Kc \subseteq L$ and therefore $(K + L)c \subseteq K$ or $(K + L)c \subseteq L$. By Corollary 3.2.14, $K + L$ is a τ-multiplication module.■

Corollary 3.3.12 *Let τ be a semicentered torsion theory on R. If K, L are τ-closed submodules of an R-module M such that $(K : L) + (L : K) \in \mathcal{F}$, then $K + L$ is a τ-multiplication module.*

Lemma 3.3.13 *Let τ be a semicentered torsion theory on R. Let N_1 and N_2 be τ-closed submodules of an R-module M. If N_1, N_2 and $N_1 + N_2$ are τ-multiplication, then $N_1 \cap N_2$ is a τ-multiplication module.*

Proof: Let $P \in K(\tau)$. If $T_P(N_1 \cap N_2) \neq N_1 \cap N_2$, then it is clear that $T_P(N_1) \neq N_1$, $T_P(N_2) \neq N_2$ and $T_P(N_1 + N_2) \neq N_1 + N_2$. By Theorem 3.3.5, there exist $x \in N_1$, $y \in N_2$, $z \in N_1 \cup N_2$, $c_1, c_2, c \in N$ such that $N_1c_1 \subseteq Cl_\tau^{N_1}(xR)$, $N_2c_2 \subseteq Cl_\tau^{N_2}(yR)$ and $(N_1 + N_2)c \subseteq Cl_\tau^{N_1+N_2}(zR)$.

Suppose $z \in N_1$ (similarly if $z \in N_2$). Then,

$$Cl_\tau^{N_1}(zR) \subseteq Cl_\tau^{N_1+N_2}(N_1) = N_1.$$

Since $y \in N_2$, $yc \in N_2$. Because $yc \in (N_1 + N_2)c \subseteq N_1$, $yc \in N_1$ because. Therefore we have

$$(N_1 \cap N_2)c_2c \subseteq Cl_\tau^{N_2}(yR)c \subseteq Cl_\tau^{N_2}(ycR).$$

On the other hand, it is obvious that $(N_1 \cap N_2)c_2c \subseteq N_1$. Thus there exists $c_2.c \in R - P$ with $(N_1 \cap N_2)cc_2 \subseteq Cl_\tau^{N_2}(ycR) \cap N_1 = Cl_\tau^{N_1 \cap N_2}(ycR)$ and by Theorem 3.2.13, $N_1 \cap N_2$ is a τ-multiplication module.■

Theorem 3.3.14 *Let τ be a semicentered torsion theory on R. Let $N_1, ..., N_k$ be τ-closed submodules of an R-module M such that N_i+N_j is a τ-multiplication module for all i, j, such that $1 \leq i < j \leq k$. Then*

1. *$N_1 + ... + N_k$ is a τ-multiplication module.*

2. *$N_1, ..., N_k$ are τ-multiplication modules if and only if $N_1 \cap ... \cap N_k$ is a τ-multiplication module.*

Proof: To prove the first part, it suffices to follow the proof of [102, Theorem 8] with slight modifications. Theorem 3.2.13 and Theorem 3.3.5 are needed.

For the second part, we use induction on k. Suppose $N_1, ..., N_k$ are τ-multiplication modules. Consider the module $X = N_2 \cap ... \cap N_k$, which is τ-multiplication. By Corollary 3.3.11, $N_1 + X$ is a τ-multiplication module and by Lemma 3.3.13, $N_1 \cap X$ is a τ-multiplication module.

Let $P \in K(\tau)$. If $T_P(N_1 + N_i) = N_1 + N_i$, then $T_P(N_1) = N_1$. Suppose that $N_1 \cap ... \cap N_k$ is a τ-multiplication module. Let $P \in K(\tau)$. Suppose $T_P(N_1 + N_i) \neq N_1 + N_i$ for all $i \in \{2, 3, ..n\}$. By Theorem 3.3.5, there exist $u_i \in N_1 \cup N_i$ and $c_i \in R - P$ such that $(N_i + N_1)c_i \subseteq Cl_\tau^{N_1+N_i}(u_i R)$. If for some i, $u_i \in N_1$, then $N_1 c_i \subseteq Cl_\tau^{N_1}(u_i R)$ and by Theorem 3.2.13, N_1 is τ-multiplication. If $u_i \in N_i$ for every $2 \leq i \leq k$. Since

$$N_1 c_i \subseteq (N_1 + N_i)c_i \subseteq N_i \ (2 \leq i \leq k),$$

we have $N_1 c_2 ... c_k \subseteq N_1 \cap ... \cap N_k$. By Corollary 3.2.14, N_1 is a τ-multiplication module. ■

Corollary 3.3.15 *Let τ a semicentered torsion theory on R. Let K_i $(1 \leq i \leq n)$ be a family of τ-closed submodules of an R-module M which are τ-multiplication modules and such that $K_i + K_j$ is τ-multiplication for $1 \leq i < j \leq n$. Then $(K_1 \cap ... \cap K_m) + (K_{m+1} \cap ... \cap K_n)$ is a τ-multiplication module for every positive integer $m < n$.*

Proof: Consider $L = K_{m+1} \cap .... \cap K_n$. By Theorem 3.3.14, L is a τ-multiplication module. By Corollary 3.3.11, $K_i + L$ is a τ-multiplication module $(1 \leq i \leq n)$ and by Corollary 3.3.11 again, $L + (K_1 \cap ... \cap K_n)$ is a τ-multiplication module.■

Now, we show what happens with the direct sum. Denote $\hat{M}_i = \oplus_{j \neq i} M_j$. Compare the next result with [31, Theorem 2.2].

Theorem 3.3.16 *Let τ be a semicentered torsion theory on R. Let M an R-module such that $M = \oplus_{i \in I} M_i$ where M_i's are τ-closed submodules of*

M. Then M is a τ-multiplication module if and only if the two following conditions are satisfied:

1. *M_i is a τ-multiplication module for each $i \in I$.*

2. *For each $i \in I$ there exists an ideal A_i of R, such that $M_i = Cl_\tau^M(M_iA_i)$ and $\hat{M}_iA_i = 0$.*

Proof: Suppose that M is a τ-multiplication module. Then $M_i \cong M/\hat{M}_i$ and therefore it is a quotient of a τ-multiplication module. Thus M_i is a τ-multiplication module.

On the other hand, since M_i is τ-closed, there exists an ideal A_i of R such that $M_i = Cl_\tau^M(MA_i) = Cl_\tau^{M_i}(MA_i)$. Thus $MA_i \subseteq M_i$. But $MA_i = (\oplus_{j\in I}M_j)A_i = \oplus_{j\in I}(M_jA_i) \subseteq M_i$. Therefore $M_j.A_i = 0$ for all $j \neq i$ and hence $\hat{M}_jA_i = 0$. Moreover, $MA_i = M_i.A_i$ and $M_i = Cl_\tau^M(M_iA_i)$.

Suppose that $P \in K(\tau)$. If $M_i = T_P(M_i)$ for all $i \in I$, then for each $m \in M_i$ there exists $c \in R - P$ such that $mc = 0$. Hence, for each $x \in M$ there exists $c \in R - P$ such that $xc = 0$. Thus $T_P(M) = M$.

Suppose that there exists $j \in I$ such that $M_j \neq T_P(M_j)$. Then by Theorem 3.2.13, there exist $c \in R - P$ and $m \in M_j$ such that $M_jc \subseteq Cl_\tau^M(mR)$. By condition 2, there exists an ideal $A_j \leq R$ such that $Cl_\tau^M(M_jA_j) = M_j$ and $\hat{M}_jA_j = 0$. We have $M_j.A_jc \subseteq Cl_\tau^M(M_jA_j)c = M_j.c \subseteq Cl_\tau^M(mR)$. If $A_j \subseteq P$, then $M_j = Cl_\tau^M(M_jA_j) \subseteq Cl_\tau^{M_j}(M_jP)$ and hence $M_j = Cl_\tau^{M_j}(M_jP)$. Therefore $M_j = T_P(M_j)$, a contradiction. Thus there exists $d \in (R-P)\cap A_j$ such that $Mcd \subseteq (\oplus_{j\in I}M_j)cd \subseteq M_jcd \subseteq Cl_\tau^M(mR)$ and by Theorem 3.2.13, M is a τ-multiplication module.■

Corollary 3.3.17 *Let τ be a semicentered torsion theory on R. Let M_i $(i \in I)$ be a family of finitely generated τ-closed modules such that $M = \oplus_{i\in I} M_i$. Then, M is a τ-multiplication module if and only if M_i is a τ-multiplication module and $ann(M_i) + ann(\hat{M}_i) \in \mathcal{F}$ for each $i \in I$.*

Proof: Suppose that M is τ-multiplication module. By Theorem 3.3.16, the first condition, M_i is a τ-multiplication module for each $i \in I$, is true. Suppose that there exists $i \in I$ such that $ann(M_i) + ann(\hat{M}_i) \notin \mathcal{F}$. Since τ is semicentered, there exists $P \in K(\tau)$ such that $ann(M_i) + ann(\hat{M}_i) \subseteq P$. From Theorem 3.3.16, there exists $A_i \leq R$ verifying $M_i = Cl_\tau^M(M_i A_i)$ and $\hat{M}_i A_i = 0$. Therefore, $A_i \subseteq ann(\hat{M}_i)$. Thus $A_i \subseteq P$. Hence $M_i = Cl_\tau^M(M_i P)$. Since M_i is finitely generated, there exists $c \in R - P$ such that $M_i c = 0$, a contradiction because $c \in ann(M_i) \cap (R - P) = \emptyset$.

Conversely, it suffices to apply Corollary 3.3.8. ■

Corollary 3.3.18 *Let τ be semicentered. Let $M = M_1 \oplus \dots \oplus M_n$ where M_i is a τ-closed τ-multiplication finitely generated module for $1 \leq i \leq n$. Then M is τ-multiplication if and only if $ann(M_i) + ann(M_j) \in \mathcal{F}$ for all $1 \leq i \neq j \leq n$.*

Proof: Suppose that M is τ-multiplication. Clearly $ann(\hat{M}_i) \subseteq ann(M_j)$, if $j \neq i$. By Corollary 3.3.17, $ann(M_i) + ann(\hat{M}_i) \in \mathcal{F}$ for $1 \leq i \leq n$. Thus $ann(M_i) + ann(M_j)$ contains an element of the Gabriel filter and therefore it belongs to the filter.

Suppose that the second part of the equivalence is true. We have $ann(\hat{M}_i) = \cap_{j\neq i} ann(M_j)$, $ann(M_i) + ann(\hat{M}_i) = ann(M_i) + \cap_{j\neq i} ann(M_j) \supseteq \Pi_{j\neq i}[ann(M_i) + ann(M_j)] \in \mathcal{F}$.■

We can apply this corollary to find all τ-multiplication modules over Dedekind domains.

Corollary 3.3.19 *τ-multiplication modules over a Dedekind domain are just modules of the form $M = Cl_\tau^M(N)$ where N is isomorphic to an ideal of R and τ-cyclic modules.*

Proof: By taking into account Proposition 3.1.32, it suffices to find finitely generated τ-multiplication modules. Let R be a Dedekind domain. A finitely generated R-module M is of the form

$$M \cong I_1 \oplus \cdots \oplus I_r \oplus R/\alpha_1 \oplus \cdots \oplus R/\alpha_n,$$

where $\alpha_i \subseteq \alpha_{i+1}$ $(1 \leq i \leq n)$, I_j $(1 \leq j \leq n)$ is an ideal of R and every R/α_i is a cyclic R-module.

Since R is a commutative noetherian ring, by [105, Corollary VI.6.15], every torsion theory is semicentered.

If τ is trivial, then every module is τ-torsion. Thus every module is τ-multiplication. Suppose that τ is not trivial. Consider three cases.

Case A : $n = 0$.

If $r = 1$, then M is a projective ideal. By [104, Theorem 1], it is multiplication.

If $r \geq 2$, then I_j is τ-closed $(1 \leq j \leq r)$ and τ-multiplication. Moreover, $ann(I_1) + ann(I_2) = 0 \notin \mathcal{F}$. By Corollary 3.3.18, M is not τ-multiplication.

Case B : $\alpha_1 \in \mathcal{F}$.

Assume $r = 0$. Then M is τ-torsion and hence it is τ-multiplication.

Suppose that $r \geq 1$. For each $x \in M$ we have

$$x \in Cl_\tau^M(I_1 \oplus \cdots \oplus I_r \oplus 0 \oplus \cdots \oplus 0).$$

Let $N = I_1 \oplus \cdots \oplus I_r$. By Theorem 3.1.19, M is τ-multiplication if and only if N is τ-multiplication.

If $r = 1$, then N is a projective ideal and therefore it is τ-multiplication. Consequently, $M = Cl_\tau^M(N)$ where N is isomorphic to an ideal of R.

If $r > 1$, then $ann(I_1) + ann(I_2) = 0 \notin \mathcal{F}$. By Corollary 3.3.18, N is not a τ-multiplication module.

Case C : $\alpha_1 \notin \mathcal{F}$.

Suppose that $n = 1$.

If $r = 0$, then $M = R/\alpha_1$ is a cyclic module and therefore it is a multiplication module.

Now, assume $r \geq 1$. We have $ann(I_1) + ann(R/\alpha_1) = \alpha_1 \notin \mathcal{F}$. By Corollary 3.3.18, M is not τ-multiplication.

Suppose $n \geq 2$.

If there exists α_k $(2 \leq k \leq n)$ such that $\alpha_k \in \mathcal{F}$, then

$$x\alpha_k \subseteq Cl_\tau^M(I_1 \oplus \cdots \oplus I_r \oplus R/\alpha_1 \oplus \cdots \oplus R/\alpha_{k-1} \oplus 0 \oplus \cdots 0).$$

Let $N = I_1 \oplus \cdots \oplus I_r \oplus R/\alpha_1 \oplus \cdots \oplus R/\alpha_{k-1} \oplus 0 \oplus \cdots 0$. By Theorem 3.1.19, M is τ-multiplication if and only if N is τ-multiplication.

If $r \geq 2$, then $ann(I_1) + ann(I_2) = 0 \notin \mathcal{F}$.

If $r = 1$, then $ann(I_1) + ann(R/\alpha_1) = \alpha_1 \notin \mathcal{F}$.

If $r = 0$ and $k - 1 = 1$, then N is cyclic and therefore it is multiplication. In this case M is τ-cyclic.

If $r = 0$ and $k - 1 \geq 2$, then $ann(R/\alpha_1) + ann(R/\alpha_2) = \alpha_2 \notin \mathcal{F}$. By Corollary 3.3.18, M is not τ-multiplication.

If none of α_j's belongs to the Gabriel filter, the situation is absolutely similar to the preceding one. ■

Immediately it follows the next corollary.

Corollary 3.3.20 *τ-multiplication modules over a P.I.D. are just τ-cyclic modules.*

Proof: Every P.I.D. is a Dedekind domain. By Corollary 3.3.19, τ-multiplication modules, for any τ, are τ-cyclic modules and the closures of ideals of the ring, which are also τ-cyclic modules in this case. ■

We shall show that in a setting where it is feasible, the product of multiplication modules relative to a same torsion theory is a relative multiplication module as well.

Proposition 3.3.21 *Let R, S be rings such that $R \subseteq S$. If X, Y are τ-multiplication R-modules inside S, then $X.Y$ is a τ-multiplication R-module.*

Proof: Let $N = Cl_\tau^{XY}(N)$. Since X and Y are τ-multiplication, $Cl_\tau^X(N \cap X) = Cl_\tau^X(XI)$ and $Cl_\tau^Y(N \cap Y) = Cl_\tau^Y(YJ)$ for some $I, J \leq R$. By applying the properties of the closure operation, we have

$$\begin{aligned} Cl_\tau^{X.Y}(N) &= Cl_\tau^{XY}(N \cap XY) \\ &= Cl_\tau^{XY}((N \cap X).(N \cap Y)) \\ &= Cl_\tau^{XY}(Cl_\tau^X(N \cap X)Cl_\tau^Y(N \cap Y)) \\ &= Cl_\tau^{XY}(Cl_\tau^X(XI)Cl_\tau^Y(YJ)) \\ &= Cl_\tau^{XY}(XYIJ). \end{aligned}$$

3.4 Relative distributive modules

Firstly, we shall study relative distributive modules, obtaining a large number of arithmetical characterizations (generalizing the ones by Năstăsescu in [90] for the case $M = R$) and showing that there are many examples. Some methods to obtain new relative distributive modules will be investigated, such as to make the tensor product or to calulate homomorphisms. It is achieved to characterize locally this kind of module and it will be check that they are narrowly related to τ-multiplication modules.

An R-module M is called *distributive* if it has distributive property of the sum with respect to the intersection or distributive property of the intersection with respect to the sum, for the lattice of submodules.

Definition 3.4.1 *A module M is called τ-distributive if the lattice of τ-closed submodules, denoted by $C_\tau(M)$, is a distributive lattice.*

Remark. This definition corresponds to that of distibutive object in the category Mod-(R, τ).

Definition 3.4.2 *A ring R is called τ-distibutive if it is distributive as an R-module.*

The case $M = R$ has been already considered in [90]. Obviously, every distributive module is a τ-distributive module for any τ. It is also immediate that every ring is τ-distributive if and only if it is τ-distributive as an R-module. If $\mathcal{F} = \{R\}$ and τ the corresponding torsion theory, then τ-distributive modules are just the distributive modules. If τ is perfect, then the R-module M is τ-distributive if and only if the R_τ-module M_τ is distributive. This is due to the isomorphism of lattices which appears in [105, Proposition 3.7].

Recall that a τ-torsionfree module M is called *τ-uniserial* if its only τ-closed submodules are M and a chain (finite or infinity) of the form

$$0 = M_0 \subset M_1 \subset \dots \subset M_n \subset \dots$$

Example 3.4.3 *Every τ-uniserial module is τ-distributive.*

In fact, if M is a τ-uniserial module and N, L, P are τ-closed submodules, then there exists a chain, for example, $N \subseteq L \subseteq P$. Then, it is clear that $(N \cap L) + (N \cap P) = N = N \cap (L + P) = N + P$.

Theorem 3.4.4 *Let τ be a semicentered torsion theory over R-Mod. For an R-module M with submodule lattice $\mathcal{L}(M)$, the following sentences are equivalent:*

1. *M is a τ-distributive R-module.*
2. *If N, L, K are submodules of M, then $Cl_\tau^M((N+L) \cap (N+K)) = Cl_\tau^M(N + (L \cap K))$.*
3. *$Cl_\tau^M(N \cap (L+N)) = Cl_\tau^M((L+N) \cap (L+K))$ for any $N, L, K \in \mathcal{L}(M)$.*
4. *M_P is distributive as an R_P-module for all $P \in K(\tau)$.*
5. *$(mR : nR) + (nR : mR) \in \mathcal{F}$ for all $m, n \in M$.*
6. *$Cl_\tau^M((m+n)R) = Cl_\tau^M((mR \cap (m+n)R) + (nR \cap (m+n)R))$ for all $m, n \in M$.*

7. $Cl_\tau^M(mR + nR) = Cl_\tau^M((m+n)R + (mR \cap nR))$ *for all* $m, n \in M$.

8. $Cl_\tau^R((K+L) : N) = Cl_\tau^R((K : N) + (L : N))$ *for all* $K, L, N \leq M$, N *being* τ*-finitely generated.*

9. $Cl_\tau^R(K : (L \cap N)) = Cl_\tau^R((K : L) + (K : N))$ *for all* $K, L, N \leq M$, L, N *being* τ*-finitely generated.*

10. $Hom_{R_P}((N/(N \cap L))_P, (L/(N \cap L))_P) = 0$ *for all* $L, N \leq M$ *and for all* $P \in K(\tau)$.

Proof: $1 \Leftrightarrow 2 \Leftrightarrow 3 \Leftrightarrow 4$ it is similar to [90, Theorem 7.3].

$1 \Rightarrow 5$. Let $m, n \in M$. By hypothesis, $Cl_\tau^R((mR : nR) + (nR : mR)) = R$. Let $P \in K(\tau)$. Then

$$Cl_{\tau_P}^{R_P}((\frac{m}{1}R_P : \frac{n}{1}R_P) + (\frac{n}{1}R_P : \frac{m}{1})R_P) = R_P.$$

By Lemma 1.1.16,

$$(R_P\frac{m}{1} : \frac{n}{1}R_P) + (\frac{n}{1}R_P : \frac{m}{1}R_P) = R_P.$$

Hence $(\frac{m}{s}R_P : \frac{n}{t}R_P) + (\frac{n}{t}R_P : \frac{m}{s}R_P) = R_P$ for all $\frac{m}{s}, \frac{n}{t} \in M_P$. By [106, Theorem 1.6], M_P is a distributive R_P-module for all $P \in K(\tau)$. By 4, M is τ-distributive.

$5 \Rightarrow 1$. Conversely, suppose that M is τ-distributive and $(nR : mR) + (mR : nR) \notin \mathcal{F}$. Since τ is semicentered, there exists $P \in K(\tau)$ such that $(nR : mR) + (mR : nR) \subseteq P$. Thus $[(mR : nR) : (nR : mR)]_P \subseteq P_P \subset R_P$. Thus M_P is distributive as an R_P-module by Lemma 3.4.4 and by [106, Theorem 1.6], $[(mR : nR) + (nR : mR)]_P = R_P$, a contradiction.

$6 \Rightarrow 5$. By using [58, Lemma 3.1], we have

$$Cl_{\tau}^{M}((m+n)R) = Cl_{\tau}^{M}((m+n)(mR:(m+n)R) + (m+n)(Rn:R(m+n)))$$
$$= Cl_{\tau}^{M}((m+n)(Rm:R(m+n)) + (Rn:R(m+n)).$$

Since $ann(m+n) \subseteq (mR:nR)+(nR:mR)$, $(mR:nR)+(nR:mR) \in \mathcal{F}$.

$1 \Rightarrow 6$ is trivial.

$7 \Rightarrow 5$. By applying [58, Lemma 3.1],

$$Cl_{\tau}^{M}(mR) = Cl_{\tau}^{M}(mR \cap (mR+nR)) = Cl_{\tau}^{M}(m((mR:nR)+(nR:mR))).$$

Since $ann(m) \subseteq (nR:mR)$, 2 follows.

$1 \Rightarrow 7$. We have $Rm \subseteq nR + (m+n)R$. Since M is τ-distributive, $Cl_{\tau}^{M}(mR) = Cl_{\tau}^{M}((nR \cap mR) + ((m+n)R \cap mR))$. Analogously, $Cl_{\tau}^{M}(nR) = Cl_{\tau}^{M}((mR \cap nR) + ((m+n)R \cap nR))$. Easily, it can be checked that

$$Cl_{\tau}^{M}(mR+nR) = Cl_{\tau}^{M}((mR \cap nR) + (m+n)R).$$

$1 \Rightarrow 8$. Let $P \in K(\tau)$. By 4, M_P is distributive as an R_P-module. By [2, Theorem 1.9], $((K_P + L_P) : N_P) = ((K_P : N_P) + (L_P : N_P))$. Since N_P is finitely generated, we have $((K+L):N)_P = ((K:N)+(L:N))_P$ for all $P \in K(\tau)$. Since τ is semicentered, 2 follows.

$8 \Rightarrow 1$. We shall prove that M_P is distributive as an R_P-module for each $P \in K(\tau)$. Let $K_P, L_P, N_P \leq M_P$, N_P being finitely generated. Since $N_P = < \frac{x_1}{1}, ..., \frac{x_r}{1} >$, there exists $N' = < x_1, ..., x_n > \leq N$ such that $(Cl_{\tau}^{M}(N'))_P = N'_P = N_P$ and obviously, $Cl_{\tau}^{M}(N')$ is τ-finitely generated. From the hypothesis, by using localization, we obtain $((K+L):N')_P = ((K:N)+(L:N))_P$. Hence $(K_P+L_P):N_P = (K_P:N_P)+(L_P:N_P)$. By [2, Theorem 1.9], M_P is distributive.

$8 \Leftrightarrow 9$. From [2, Theorem 1.9], it suffices to use localization.

$1 \Leftrightarrow 10$ is straightforward by applying [106, Proposition 1.1]. ■

Corollary 3.4.5 *Let τ be a semicentered torsion theory on R. Let M be a τ-distributive R-module. Let L, N be submodules of M. Then*

1. *if N is finitely generated and $N \cap L = 0$, then $Hom_R(N, L)$ is τ-torsion.*
2. *if M/L is finitely generated and $N + L = M$, then the R-module $Hom_R(M/L, M/N)$ is τ-torsion.*

Proof: We shall prove 1. Let $P \in K(\tau)$. By Theorem 3.4.4 (10), we have that $Hom_{R_P}(N_P, L_P) = 0$. Since N is finitely generated, the canonical morphism $(Hom_R(N, L))_P \longrightarrow Hom_{R_P}(N_P, L_P)$ is injective. It is deduced that $(Hom_R(N, L))_P = 0$ for all $P \in K(\tau)$. Hence 1 follows.

Now, we prove 2. Let $P \in K(\tau)$. By Theorem 3.4.4 (10),

$$Hom_{R_P}((N/(N \cap L))_P, (L/(N \cap L))_P) = 0 \cong$$
$$Hom_{R_P}(((N + L)/L)_P, ((N + L)/N)_P) =$$
$$Hom_{R_P}((M/L)_P, (M/N)_P).$$

Since M/L is finitely generated, $(Hom_R(M/L, M/N))_P = 0$ for all $P \in K(\tau)$. Therefore, the R-module $Hom_R(M/L, M/N)$ is τ-torsion.■

Example 3.4.6 Let τ be a semicentered torsion theory. By Theorem 3.4.4 (4), every Krull domain R is a τ-distributive ring, i.e., it is τ-distributive as an R-module.

The next proposition gives more examples of τ-distributive modules.

Proposition 3.4.7 *If τ is semicentered, then every submodule and every quotient of a τ-distributive module is a τ-distributive module.*

Proof: Suppose that the R-module M is τ-distributive and let N be a submodule of M. Let $P \in K(\tau)$. Then, N_P is an R_P-submodule of M_P that is distributive by Theorem 3.4.4 (4). Thus N_P is distributive as an R_P-module. Since τ is semicentered, by Theorem 3.4.4 (4) again, it follows that N is τ-distributive.

To prove that M/N is τ-distributive, it is not necessary that τ is semicentered. It suffices to use the definition of τ-distributive module and to apply that the τ-closed submodules in M/N are of the form L/N where L is a τ-closed in M containing N. ■

Proposition 3.4.8 *Let τ be a semicentered torsion theory over R-Mod. If $M = Cl_\tau^M(N)$, then M is τ-distributive if and only if N is τ-distributive.*

Proof: Suppose that M is τ-distributive. By Theorem 3.4.4 (4), M_P is a distributive R_P-module for all $P \in K(\tau)$. Then $M_P = (Cl_\tau^M(N))_P = Cl_{\tau_P}^{M_P}(N_P)$ by Lemma 1.1.16. Moreover, $M_P = N_P$. Thus N_P is distributive as an R_P-module. By Theorem 3.4.4 (4) again, N is τ-distributive. The converse can be proved in the same way.■

Definition 3.4.9 *A module is called τ-Bezout if every τ-finitely generated submodule is τ-cyclic.*

Proposition 3.4.10 *τ-distributive modules over a P.I.D. are just τ-Bezout modules.*

Proof: Straightforward from Corollary 3.3.20.■

The following results give different ways to obtain new relative distributive modules from relative distributive modules .

Proposition 3.4.11 *Let τ be a semicentered torsion theory on R. Let M, N be two τ-distributive R-modules. Then*

1. *$M \otimes_R N$ is a τ-distributive R-module.*

2. *If M is finitely generated, then $Hom_R(M, N)$ is a τ-distributive R-module.*

Proof: By Theorem 3.4.4 (4), 1 is trivial.

Since M, N are τ-distributive modules, M_P, N_P are distributive R_P-modules for all $P \in K(\tau)$. By [2, Lemma 4.1], $Hom_{R_P}(M_P, N_P)$ is distributive as an R_P-module. Since M is a finitely generated R-module, the canonical morphism $[Hom_R(M, N)]_P \longrightarrow Hom_{R_P}(M_P, N_P)$ is injective. By Proposition 3.4.7, $[Hom_R(M, N)]_P$ is distributive for all $P \in K(\tau)$. By Theorem 3.4.4 (4), $Hom_R(M, N)$ is a τ-distributive R-module.■

Remark 3.4.12 If M is τ-finitely generated and N is τ-torsion free the same result is obtained. It suffices to realize that with the above hypothesis if $M = Cl_\tau^M(F)$, then two maps belonging to $[Hom_R(M, N)]_P$ which are equal over F are equal over M.

For an R-module M, set $\tau - Supp(M) = \{P \in K(\tau); M_P \neq 0\}$.

Proposition 3.4.13 *Let M_i $(i \in I)$ be a family of τ-distributive modules. Then $\oplus_{i \in I} M_i$ is τ-distributive if and only if $\tau - Supp(M_i) \cap \tau - Supp(M_j) = \emptyset$ for all $i, j \in I$ $i \neq j$.*

Proof: Suppose that $\oplus_{i \in I} M_i$ is τ-distributive and for some $i \neq j$, there exists $P \in K(\tau)$ such that $(M_i)_P \neq 0 \neq (M_j)_P$. The R_P-module $(\oplus_{i \in I}(M_i))_P \cong$

$\oplus_{i\in I}(M_i)_P$ is distributive. By [2, Proposition 1.8], $Supp(M_i) \cap Supp(M_j) = \emptyset$. However $P \in Supp(M_i) \cap Supp(M_j) = \emptyset$, a contradiction.

Let $P \in K(\tau)$. If there exists $Q.R_P \in Supp((M_i)_P) \cap Supp((M_j)_P)$, then, since $(M_i)_{P_{Q.R_P}} \cong M_Q$, $(M_i)_Q \neq 0 \neq (M_j)_Q$ for $i \neq j$. If $Q \in \mathcal{F}$, then $P \in \mathcal{F}$ as $Q \subseteq P$. Therefore, $Q \in K(\tau)$, a contradiction. ■

The following theorem establishes a relationship between τ-distributive modules and τ-multiplication modules. It is a generalization of [13, Proposition 7].

Theorem 3.4.14 *Let τ be a semicentered torsion theory on $R-Mod$. Then M is τ-distributive if and only if every τ-finitely generated submodule of M is τ-multiplication.*

Proof: Suppose that M is τ-distributive. Let $N = Cl_\tau^N(F) \leq M$, F being finitely generated. By Theorem 3.4.4 (4), M_P is distributive for all $P \in K(\tau)$. Since F_P is finitely generated for all $P \in K(\tau)$ as an R_P-module, by [13, Proposition 7], the R_P-module F_P is multiplication for all $P \in K(\tau)$. By Theorem 3.2.5, F is a τ-multiplication R-module. By Theorem refparche, N is τ-multiplication as an R-module.

Conversely, suppose that every τ-finitely generated submodule of M is τ-multiplication. Let $P \in K(\tau)$. We shall prove that M_P is distributive as an R_P-module. Let $N_P =< \frac{x_1}{1}, ..., \frac{x_r}{1} >$ with $x_i \in N$ $(1 \leq i \leq r)$. Let $L_P \leq N_P$. Consider $K =< x_1, ..., x_r >\leq N$. Obviously, $K_P = N_P$. Since $Cl_\tau^N(K) \leq M$ is τ-finitely generated, it is τ-multiplication. Hence $Cl_\tau^N(L \cap N) = Cl_\tau^N(Cl_\tau^N(K)I)$ for some ideal I of R. By localization, $L_P = N_P.I_P$. Thus N_P is multiplication as an R_P-module. ■

Corollary 3.4.15 *Let τ be a semicentered torsion theory. Every τ-noetherian τ-distributive module is a strongly τ-multiplication module.*

Corollary 3.4.16 *If τ is semicentered, then every τ-uniserial τ-noetherian module M is strongly τ-multiplication.*

Corollary 3.4.17 *Let τ be a semicentered torsion theory. Every τ-noetherian τ-distributive ring is a τ-multiplication ring.*

In particular, if τ is the canonical torsion theory, then every Krull domain is a τ-multiplication ring, by Corollary 3.4.17.

The following example shows that the τ-distributive modules class is strictly wider than the distributive modules class.

Example 3.4.18 Let $R = K[x, y]$, K being a field. Let τ be the canonical torsion theory. Since R is a Krull domain, it is a τ-multiplication ring. Thus every submodule of K is τ-multiplication. By Theorem 3.4.14, R is a τ-distributive R-module.

R is a integral domain which is not a Dedekind domain. By [69, Proposition 9.13], there exists some ideal (which must be finitely generated as R is noetherian) which is not multiplication as an R-module. By [13, Proposition 7], R is not distributive as an R-module.

The following results study the endomorphism ring of τ-distributive modules.

Proposition 3.4.19 *If $M = M_1 \oplus M_2$ is a τ-distributive τ-torsion free R-module, then $Hom_R(M_1, M_2) = 0$.*

Proof: First, we prove that M_1 and M_2 are τ-closed. We check it for M_1. Let $x = (m_1, m_2) \in M$. If $x \in Cl_\tau^M(M_1)$, then there exists an ideal $I \in \mathcal{F}$ such that $xI \subseteq M_1$. Therefore, $m_2 I = 0$ and $(0, m_2) \in Cl_\tau^M(0) = \tau(M) = 0$. It follows that $m_2 = 0$ and hence, $x \in M_1$. Analogously, it is deduced that $M_2 = Cl_\tau^M(M_2)$. Let $f \in Hom_R(M_1, M_2)$ and let γ its graph, i.e., $\gamma = \{(a, f(a); a \in M_1\}$. We take $x = (m_1, m_2) \in Cl_\tau^M(\gamma)$. Then, there exists an ideal $I \in \mathcal{F}$ such that $xI \in \gamma$. Thus $m_2 I = f(m_1)I$. Hence, $(m_1 - f(m_2)I) = 0$ and therefore, $m_2 - f(m_1) \in \tau(M) = 0$. As a result, $x \in \gamma$ and γ is also τ-closed. Since M is τ-distributive, its lattice of τ-closed submodules is distributive and as a consequence,

$$\begin{aligned}\gamma = \gamma \cap M &= \gamma \cap (M_1 \cap M_2)\\ &= (\gamma \cap M_1) + (\gamma \cap M_2)\\ &= \gamma \cap M_1 \subseteq M_1.\end{aligned}$$

It follows that $f(x) = 0$ for every $x \in M_1$ and hence, $f = 0$. ■

In the next results, we suppose always that τ is semicentered.

Proposition 3.4.20 *If R is τ-distributive and M is τ-multiplication, then M is τ-distributive.*

Proof: Let $P \in K(\tau)$. Then, R_P is distributive by [90, Theorem 7.3], M_P is multiplication as an R_P-module by Theorem 3.2.5. By [35, Proposition 1.2], M_P is a distributive R_P-module for every $P \in K(\tau)$. By Theorem 3.4.4, M is τ-distributive as an R-module. ■

Theorem 3.4.21 *If M is τ-finitely generated and faithful, then M is τ-distributive if and only if M is τ-multiplication and R is τ-distributive.*

Proof: Suppose that M is τ-distributive and let $P \in K(\tau)$. By Theorem 3.4.4 (4), M_P is distributive as an R_P-module. It is also finitely generated as M is τ-finitely generated. By Lemma 1.1.16, it can be applied [35, Theorem 1.3] to follow that M_P is a multiplication R_P-module and that R_P is a distributive ring. By [90, Theorem 7.3], R is τ-distributive and by Theorem 3.2.5, M is τ-multiplication. The other implication is immediate by Proposition 3.4.20. ■

Proposition 3.4.22 *If M is a faithful τ-finitely generated τ-distributive R-module, then $M_P \cong R_P$ for every $P \in K(\tau)$ and R is τ-distributive.*

Proof: Let $P \in K(\tau)$. By Theorem 3.4.4 (4), M_P is distributive as an R_P-module. By [2, Proposition 1.3], the lattice of R_P-submodules of M_P is linearly ordered. Since M_P is finitely generated, it is cyclic, i.e., $M_P \cong (R_P/PR_P)$. By Lemma 1.1.16, $PR_P = ann_{R_P}(M_P) = 0$ and it follows that $M_P \cong R_P$. Since every R_P for $P \in K(\tau)$ is distributive, by [90, Theorem 7.3], R is τ-distributive. ■

3.5 Relative multiplication ideals and rings

We shall characterize multiplication ideals relative to a torsion theory in terms of its behavior with respect to τ-critical ideals by associating them to certain ideals. In case of rings, completely integrally closed domains, between others, are proved to be multiplication domains for a certain torsion theory.

The following results give arithmetical characterizations of τ-multiplication ideals.

Lemma 3.5.1 *Let I be an ideal of R. Then I is τ-multiplication (with the induced torsion) if and only if $I \cap B = Cl_\tau^I((B:I).I)$ for all τ-closed $B \leq R$.*

Proof: We suppose the right part of the equivalence is true. Let $J \leq I$ such that J is τ-closed in I. By the hypothesis,

$$Cl_\tau^R(J) \cap I = Cl_\tau^I((Cl_\tau^R(J):I).I).$$

But $Cl_\tau^R(J) \cap I = Cl_\tau^I(J) = J$, so I is multiplication.

We suppose the left part of the equivalence is true. Then let B be such that $B \leq R$ and $Cl_\tau^R(B) = B$. It is clear that $(I \cap B : I) = (B : I)$. By Lemma 2.2.18 and as $Cl_\tau^I(B \cap I) = B \cap I$, then $I \cap B = Cl_\tau^I((B:I).I)$.■

Definition 3.5.2 *A τ-closed ideal I of R is said to be a τ-meet-principal ideal if $Cl_\tau^R(AI \cap B) = Cl_\tau^R([A \cap (B:I)]I)$ for any τ-closed ideals A, B.*

Proposition 3.5.3 *The ideal $I = Cl_\tau^R(I)$ is τ-multiplication if and only if it is τ-meet-principal.*

Proof: First, suppose that I is τ-multiplication and let A, B be τ-closed ideals of R. Then, we have

$$Cl_\tau^R(AI \cap B) = Cl_\tau^R([(AI \cap B):I]I)$$
$$= Cl_\tau^R([(AI:I) \cap (B:I)]I) = Cl_\tau^R((AI:I)I \cap (B:I)I)$$
$$= Cl_\tau^R(Cl_\tau^R((AI:I)I) \cap Cl_\tau^R((B:I)I)) = Cl_\tau^R(Cl_\tau^R(AI)Cl_\tau^R((B:I)I))$$
$$= Cl_\tau^R(AI \cap (B:I)I) = Cl_\tau^R([A \cap (B:I)]I).$$

■

Corollary 3.5.4 *If I, J are τ-multiplication τ-closed ideals, then $Cl_\tau^R(IJ)$ is τ-multiplication.*

Proof: Let A, B τ-closed ideals of R. By Proposition 3.5.3, it suffices to show that $Cl_\tau^R(IJ)$ is τ-meet-principal. By using that I and J are τ-meet-principal, we have

$$\begin{aligned} Cl_\tau^R(AIJ \cap B) &= Cl_\tau^R([AI \cap (B:J)]J) \\ &= Cl_\tau^R(([A \cap ((B:J):I)]I)J) \\ &= Cl_\tau^R([A \cap (B:IJ)]IJ). \end{aligned}$$

■

Lemma 3.5.5 *If R is a τ-multiplication ring and $P \in Max_\tau(R)$, then there is no τ-closed ideals properly contained between P and $Cl_\tau^R(P^2)$.*

Proof: Suppose that $A = Cl_\tau^R(A) \leq R$ and $Cl_\tau^R(P^2) \subsetneq A \subsetneq P$. Since P is a τ-multiplication ideal, there exists $B \leq R$, which can be taken τ-closed, such that $A = Cl_\tau^R(PB)$. If $B \subsetneq P$, then $Cl_\tau^R(B+P) = R$. By multiplying by P, we have $PCl_\tau^R(B+P) = M$. By taking closures, we obtain $P = Cl_\tau^R(Cl_\tau^R(P)Cl_\tau^R(B+P)) = Cl_\tau^R(PB+P)$. Since $Cl_\tau^R(PB+P^2) = Cl_\tau^R(Cl_\tau^R(PB)+Cl_\tau^R(P^2))$, it follows that $P = Cl_\tau^R(A+Cl_\tau^R(P^2)) = A$. ■

If $I = Cl_\tau^R(I)$, we consider the ideal $\eta(I) = \sum_{x\in I}(Cl_\tau^R(Rx) : I)$ that clearly contains I.

Lemma 3.5.6 *If I is a τ-closed τ-multiplication ideal, then for every ideal $A = Cl_\tau^R(A) \subseteq I$, we have $A = Cl_\tau^R(A\eta(I))$*

Proof: Take $x \in I$. Since I is τ-multiplication, $Cl_\tau^R(Rx) = Cl_\tau^R(I(Rx:I))$. Therefore,

$$I = \textstyle\sum_{x\in I} Cl_\tau^R(Rx) = Cl_\tau^R(\sum_{x\in I} Cl_\tau^R(I(Cl_\tau^R(Rx):I))$$

$$= Cl_\tau^R(\textstyle\sum_{x\in I} I(Cl_\tau^R(Rx) : I)) = I\eta(I).$$

Since I is τ-multiplication, $A = Cl_\tau^R(BI)$ for some ideal $B = Cl_\tau^R(B)$ of R. Then, $A = Cl_\tau^R(BCl_\tau^R(I\eta(I))) = Cl_\tau^R(BI\eta(I)) = Cl_\tau^R(Cl_\tau^R(BI)\eta(I)) = Cl_\tau^R(A\eta(I))$. ■

Lemma 3.5.7 *Given $A, B \in C_\tau(R)$, it holds*

i) $\eta(A)\eta(B) \subseteq \eta(AB)$.

ii) $\eta(I)_S \subseteq \eta(I_S)$ *for every multiplicatively closed subset S of R and for every ideal I of R.*

Proof: *i)* Let $a \in A$, $b \in B$, $x \in (Cl_\tau^R(Ra : A)$ and $y \in (Cl_\tau^R(Rb) : B)$. Then, $Ax \subseteq Cl_\tau^R(Ra)$, $By \subseteq Cl_\tau^R(Rb)$. By multiplying, $xyAB \subseteq Cl_\tau^R(Ra)Cl_\tau^R(Rb) \subseteq Cl_\tau^R(Rab)$. Hence $(Cl_\tau^R(Ra) : A)(Cl_\tau^R(Rb) : B) \subseteq (Cl_\tau^R(Rab) : AB)$. By the definition of η, *i)* follows.

ii) If $y \in (Cl_\tau^R(Rx) : I)_S$ for $x \in I$, then $y = \frac{a}{s}$ with $a \in Cl_\tau^R(Rx)$, that is, there exists $J \in \mathcal{F}$ such that $aJ \subseteq Rx$. Then, it is easy to check that $y \in (Cl_\tau^R(R_S\frac{x}{1}) : I_S)$. Now, we have $\eta(I)_S = (\sum_{x\in I}(Cl_\tau^R(Rx) : I))_S = \sum_{x\in I}(Cl_\tau^R(Rx) : I)_S \subseteq \sum_{x\in I}(Cl_\tau^R(R_S\frac{x}{1}) : I_S) \subseteq \eta(I_S)$. ■

We need to work with torsion theories τ having sufficient τ-criticals. Therefore, now τ is a such torsion theory.

Proposition 3.5.8 *Assume that $P \in Max_\tau(R)$ and $I \in C_\tau(R)$ is τ-multiplication. If $\eta(I) \subseteq P$, then $I_P = 0$ and if $\eta(I) \nsubseteq \eta(I)$, then I_P is principal.*

Proof: Suppose that $\eta(I) \subseteq P$ and take $x \in I$. By Lemma 3.5.6, $Cl_\tau^R(Rx) = Cl_\tau^R(Rx\eta(I))$. Therefore, there exists $J \in \mathcal{F}$ such that $Jx \subseteq \eta(I)$. Thus there exists $c \in J \cap (R - P)$ such that $cx = 0$. It follows that $I_P = 0$. Since I

is τ-multiplication and P is a τ-closed prime, I_P is multiplication as an R_P-module by Theorem 3.2.5. But every multiplication module over a local ring is principal (cf. [13]). ■

Proposition 3.5.9 *Every τ-closed ideal I is τ-multiplication if and only if for every $P \in Max_\tau(R)$ such that $\eta(I) \subseteq P$, it holds $I_P = 0$.*

Proof: The necessity is a consequence from Proposition 3.5.8.

In order to prove the sufficiency, we see that I is τ-meet-principal. Given $A, B \in C_\tau(R)$, the inclusion $Cl_\tau^R([A \cap (B : I)]I) \subseteq Cl_\tau^R(AI \cap B)$ is always true. Let $P \in Max_\tau(R)$. First, suppose that $\eta(I) \subseteq P$. To check the equality, we localize in the τ-closed maximals. Then, $[Cl_\tau^R(A \cap (B : I)I)]_P \subseteq Cl_\tau^R(A_P \cap (B : I)_P I_P) = Cl_\tau^R(0)$. Therefore, $Cl_\tau^R(A \cap (B : I)I) = Cl_\tau^R(0)$. On the other side, $[Cl_\tau^R(AI \cap B)]_P \subseteq Cl_\tau^R(A_P I_P \cap B_P) = Cl_\tau^R(0)$. It follows that $[Cl_\tau^R(AI \cap B)] = Cl_\tau^R(0)$.

Now, we suppose that $\eta(I) \not\subseteq P$. By definition of $\eta(I)$, there exists $x \in I$ such that $(Cl_\tau^R(Rx) : I) \not\subseteq P$. Then there exists $t' \in R - P$ verifying $t'I \subseteq Cl_\tau^R(Rx)$, that is, there is $J \in \mathcal{F}$ with $t'JI \subseteq Rx$. But some element of J does not belong to P because otherwise $P \in \mathcal{F}$. Thus there exists $t \in R - P$ such that $tI \subseteq Rx$. Now, we can proceed as in the absolute case. Given $b = a_1 i_1 + \cdots a_n i_n \in AI \cap B$ with $a_j \in A$, $i_j \in I$ $(1 \leq j \leq n)$, we have $ti_j \in tI \subseteq Rx$. Consequently, for every j, $ti_j = r_j x$ for some $r_j \in R$. Hence $tb = a_1(ti_1) + \cdots + a_n(ti_n) = a_1(r_1 x) + \cdots + a_n(r_n x) = (a_1 r_1 + \cdots + a_n r_n)x$. Then, $(a_1 r_1 + \cdots + a_n r_n)tI \subseteq (a_1 r_1 + \cdots + a_n r_n)Rx = Rtb \subseteq B$. Therefore, $(a_1 r_1 + \cdots a_n r_n)t \in A \cap (B : I)$. Hence $\frac{a_1 r_1 + \cdots + a_n r_n}{1} \in [A : (B \cap I)]_P$. Then, $\frac{b}{1} = (\frac{t}{1})^{-1} \frac{tb}{1} = (\frac{t}{1})^{-1} \frac{a_1 r_1 + \cdots + a_n r_n}{1} \frac{x}{1} \in A_P \cap (B : I)_P I_P$. We have

$[Cl_\tau^R(AI \cap B)] = (AI \cap B)_P \subseteq [A \cap (B : I)]_P I_P \subseteq Cl_\tau^R([A \cap (B : I)]_P I_P)$. This gives the equality. ■

Proposition 3.5.10 *If I is a τ-closed τ-locally principal ideal of R, then the ideal $\eta(I)$ belongs to the Gabriel filter.*

Proof: Let $P \in Max_\tau(R)$. By hypothesis, I_P is principal. It follows that there is $x \in I$ such that $I_P = (Rx)_P$. Hence $R_P = ((Rx)_P : I_P) = (Rx : I)_P = (Cl_\tau^R(Rx) : I)_P$ because I is τ-finitely generated. Then, $\eta(I)_P = [\sum_{x \in I}(Cl_\tau^R((Rx) : I)]_P = [\sum_{x \in I}(Cl_\tau^R(Rx : I)]_P = R_P$. As a result, $Cl_\tau^R(\eta(I)) = R$. ■

Proposition 3.5.11 *Let $I \in C_\tau(R)$. Then*

i) If $Cl_\tau^R(\eta(I)) = R$, then I is τ-locally principal.

ii) If $\eta(I) = R$, then I is τ-finitely generated.

Proof: Let $P \in Max_\tau(R)$. Since $\eta(I) = \sum_{x \in I}(Cl_\tau^R(Rx) : I)$ and $\eta(I) = R$, it follows that there exists $x \in I$ such that $(Cl_\tau^R(Rx) : I) \nsubseteq P$. Thus there is $t' \in R - P$ such that $tI \subseteq tI \subseteq Rx$. As a result, $I_P = (Rx)_P$. Hence I is τ-locally principal.

Now we suppose that $\eta(I) = R$. Then, there exist $x_1, \cdots, x_n \in I$ such that $(Cl_\tau^R(Rx_1) : I) + \cdots + (Cl_\tau^R(Rx_n) : I) = R$. By multiplying by I, $I = (Cl_\tau^R(Rx_1) : I)I + \cdots + (Cl_\tau^R(Rx_n) : I)I \subseteq Cl_\tau^R(Rx_1 + \cdots Rx_n) \subseteq I$. Therefore, I is τ-finitely generated. ■

Theorem 3.5.12 *Given $P \in Max_\tau(R)$, P is τ-multiplication if and only if either P is τ-finitely generated and τ-locally principal or P is τ-locally principal and $P \subset \eta(I) \subset Cl_\tau^R(\eta(P)) = R$ or R_P is a field.*

Proof: First, notice that by the maximality of P, there are two possibilities, namely, $P = Cl_\tau^R(\eta(P))$ or $Cl_\tau^R(\eta(P)) = R$. If $P = Cl_\tau^R(\eta(P))$, then $\eta(P) = P$ and by Proposition 3.5.9, P is τ-multiplication if and only if $P_P = 0$, that is, R_P is a field.

Suppose that P is τ-multiplication. If $Cl_\tau^R(\eta(P)) = R$, by Proposition 3.5.11, P is always τ-locally principal. If $\eta(P) = R$, the same proposition gives that P is τ-finitely generated. Otherwise, $\eta(P) \neq R$ but $Cl_\tau^R(\eta(P)) = R$, that is, $P \subset \eta(P) \subset Cl_\tau^R(\eta(P)) = R$.

Now we suppose that the right side of the equivalence is true. If P is τ-finitely generated, then being τ-locally principal is equivalent to being τ-multiplication by Theorems 3.1.19 and 3.2.5. If $P \subset \eta(P) \subset Cl_\tau^R(\eta(P)) = R$, then there is no $Q \in Max_\tau(R)$ such that $\eta(P) \subseteq Q$ and by Proposition 3.5.9, P is τ-multiplication. ∎

Lemma 3.5.13 *If $A \in C_\tau(R)$, $ann(A)$ is contained in finitely many ideals belonging to $Max_\tau(R)$, say $P_1, ..., P_n$ and A_{P_i} is principal in R_{P_i} $(1 \leq i \leq n)$, then A is τ-principal in R.*

Proof: By hypothesis, $A_{P_i} = (Rx_i)_P$ with $x_i \in A$ for each $1 \leq i \leq n$. Let $P \in Max_\tau(R)$. Since $ann(A) \not\subseteq P$, then $A_P = 0$. Therefore, $A_P = < x_1, ..., x_n >_P$ for each $P \in Max_\tau(R)$. We choose $y_i \in \cap_{j \neq i} M_j - M_i$ and we take $z = y_1x_1 + ... + y_nx_n$. Then, for every $P \in Max_\tau(R)$, we have $A_P = (Rz)_P$ and therefore, $A = Cl_\tau^R(Rz)$. ∎

Proposition 3.5.14 *If $Max_\tau(R)$ is a finite set and $A \in C_\tau(R)$, the following sentences are equivalent:*

1. *A is τ-multiplication.*

2. A is τ-principal.

3. A is τ-locally principal.

Proof: $1 \Leftarrow 3$. If A is τ-multiplication, then for every $P \in Max_\tau(R)$, A_P is principal always. Then A is τ-locally principal.

$3 \Leftarrow 2$. From Lemma 3.5.13. $2 \Leftarrow 1$. Because every τ-cyclic module is τ-multiplication by Corollary 3.1.21. ■

Lemma 3.5.15 *If $I = Cl^R_\tau(I)$ is τ-multiplication with $rank(I) > 0$, then I is τ-finitely generated or $I \subset \eta(I) \subseteq Cl^R_\tau(\eta(I)) = R$.*

Proof: Assume that I is not τ-finitely generated. Then, by Proposition 3.5.11, $\eta(I) \neq R$. If moreover, $Cl^R_\tau(\eta(I)) = R$, then there exists $P \in Max_\tau(R)$ such that $\eta(I) \subseteq Cl^R_\tau(\eta(I)) \subseteq P$. By Proposition 3.5.8, $I_P = 0$ and $rank(I) \leq rank(I_P) = 0$, a contradiction. The other alternative is $Cl^R_\tau(\eta(I)) = R$. Moreover, $I \neq \eta(I)$ because in that case, $I = Cl^R_\tau(\eta(I))$. ■

The next result is relative version of Mott's Theorem, which establishes that if every prime ideal is multiplication, then the ring is multiplication.

Theorem 3.5.16 *If R is τ-noetherian and every τ-closed prime is multiplication, then R is τ-multiplication.*

Proof: Since R is τ-noetherian, R is τ-multiplication if and only if the rings R_P are multiplication for every $P \in Spec_\tau(R)$. By Mott's Theorem this happens if and only if every prime ideal of R_P is multiplication as an R_P-module. A prime ideal of R_P is of the form Q_P with $Q \in Spec(R)$ but Q can be taken from the τ-closed prime of R as $Cl^R_\tau(Q)_P = Q_P$. Since every

$Q \in Spec_\tau(R)$ is multiplication, its localized one is multiplication as well and theherefore the proof ends. ■

Now, we shall study relative multiplication domains. The concepts of star-multiplication domain and τ-multiplication domain are related and completely integrally closed domains are characterized in terms of torsion theories. Then, we deal with relative multiplication rings in general. Finally, divisorially multiplication rings are defined and investigated.

Now, R stands for an integral domain. Let τ be a hereditary torsion theory on R. Let K be the field of quotients of R. In the absolute case, multiplication domains are just Dedekind domains. In the relative case, we have the following result.

Definition 3.5.17 *A τ-Dedekind domain R is an integral domain in which every non zero τ-closed ideal is τ-invertible.*

Theorem 3.5.18 *Multiplication domains with respect to τ are exactly τ-Dedekind domains.*

Proof: Suppose that R is a τ-Dedekind domain. Let A be a τ-closed ideal of R. By hypothesis, A is τ-invertible. By Example 3.1.13, A es τ-multiplication and, hence so is R.

Now, suppose that R is a τ-multiplication domain. Let $0 \neq A$ a τ-closed ideal of R. Take $0 \neq a \in A$. Since R is a domain, there exists a fractional ideal C of R such that $aR.C = R$. By hypothesis, $Cl^R_\tau(aR) = Cl^R_\tau(B.A)$ for some $B \leq R$. It holds $R = Cl^R_\tau(aR.C) = Cl^R_\tau(B.A.C)$ and hence, A is τ-invertible. It follows that R is a τ-Dedekind domain. ■

Remark. Every Krull domain is a τ-Dedekind domain, τ being the canonical torsion theory (see Example 3.1.8).

Recall that an R-module M is called *τ-cyclic* if $M = Cl_\tau^M(Rm)$ for some $m \in M$. Now, we are interesting in knowing when a τ-multiplication module is τ-cyclic. Let R be an integral domain and let K be its field of quotients. We denote by $E(M)$ the injective hull of an R-module M. The classical torsion theory on R, μ, is defined in the following way: if M is an R-module, its torsion submodule is

$$\mu(M) = \{m \in M; \exists d \in R - \{0\} \wedge md = 0\}.$$

Theorem 3.5.19 *Let M be a non zero R-module. Then, the following statements are equivalent:*

1. *M is μ-multiplication.*
2. *M is μ-cyclic.*
3. *There exists a monomorphism of R-modules from M into K.*

Proof: We prove the equivalence between 1 and 2. Every τ-cyclic module is τ-multiplication by Corollary 3.1.21.

Now, suppose that M is μ-multiplication. Firstly, we assume that $\mu(M) = 0$. Let $0 \neq m \in M$. By hypothesis, there exists $A \leq R$ such that $Cl_\mu^M(mR) = Cl_\mu^M(M.A)$. Moreover, $A \neq 0$ because if not, $m \in \mu(M)$, a contradiction. Given $n+mR \in M/mR$, there exists a non zero element $a \in A$ such that $na \in Cl_\mu^M(mR)$. Therefore, there exists $I \in \mathcal{F}(\mu)$ such that $naI \subseteq mR$. It follows that there exists $0 \neq i \in I$ verifying that $nai \in mR$. Thus $(n + mR)ai \in$

M/mR. Consequently, M/mR is μ-torsion and hence, $M = Cl_\mu^M(mR)$. Consider the general case. Then, $M/\mu(M)$ is μ-torsion free. By the above case, there exists $m \in M$ such that $M/\mu(M) = Cl_\mu^{M/\mu(M)}(m + \mu(M))$. As a consequence, for every $x \in M - \mu(M)$, there exists a non zero element, $a \in R$, such that $xa + \mu(M) = m + \mu(M)$. It is deduced that M/mR is μ-torsion and hence, M is μ-cyclic.

Prove that 2 is equivalent to 3. Suppose that $M = Cl_\mu^M(mR)$ for some $m \in M$. Then, M/mR is μ-torsion and mR is essential in M. Let $0 \neq x \in M$. Then, $0 \neq xr$ for some $0 \neq r \in R$ and $mR \cap rR \neq 0$. Since $mR \cong R$, it holds $E(M) \cong E(Rm) \cong K$. It is deduced that $2 \Rightarrow 3$.

Suppose that $\varphi : M \to K$ is a monomorphism of R-modules. Since $M \neq 0$, $\varphi(M) \cap R \neq 0$. Let $0 \neq a \in \varphi(M) \cap R$. It is clear that K/aR is μ-torsion. Since μ is hereditary, $\varphi(M)/aR$ is μ-torsion. Let $m \in M$ such that $a = \varphi(m)$. It follows that M/mR is μ-torsion and hence M is μ-cyclic.
■

As every ideal of any integral domain verifies condition 3 of Therorem 3.5.19, the following result is straightforward.

Corollary 3.5.20 *Every integral domain is μ-multiplication.*

Recall that a *star-operation* over a domain R with field of quotients K is a mapping, $*$, from the set of all fractional ideals of R, $Frac(R)$, into itself verifying the following properties:

1. $(aR)^* = aR$ for every $0 \neq a \in K$.

2. $(aI)^* = aI^*$ for every $0 \neq a \in K$ and for every $I \in Frac(R)$.

3. $I \subseteq J \Rightarrow I^* \subseteq J^*$ for every $I, J \in Frac(R)$.

4. $I \subseteq I^*$ and $(I^*)^* = I^*$ for every $I \in Frac(R)$.

A $*$-*ideal*, I, is a fractional ideal verifying that $I = I^*$. A domain R is $*$-*multiplication* if for every $*$-ideal A there exists a fractional ideal, B, such that $(B.A)^* = R$.

If τ is a hereditary torsion theory on the domain R such that R is τ-closed in its field of quotients, K, then the mapping $*$ defined by $I^* = Cl_\tau^K(I)$ is a star-operation in R (see [47]). In this situation, we have the following result.

Proposition 3.5.21 *R is a τ-multiplication ring if and only if R is a $*$-multiplication ring.*

Proof: Suppose that R is τ-multiplication. Let $A \leq R$ be a $*$-ideal. By hypothesis and Theorem 3.5.18, there exists $B \in Frac(R)$ such that $R = Cl_\tau^R(B.A)$. Thus $(B.A)^* = R$ and B^* is a $*$-ideal verifying that $(B^*.A)^* = Cl_\tau^R(B^*.A) = R$. The result can be extended easily for fractional $*$-ideals.

Conversely, we assume that R is $*$-multiplication. Let A be a τ-closed ideal of R. By hypothesis, there exists $B = B^*$ such that $(B.A)^* = R = Cl_\tau^R(B.A)$. Therefore, A is τ-invertible. By Theorem 3.5.18, A is τ-multiplication and hence, so is R. ■

On the other hand, $\mathcal{F} = \{I \leq D; I^* = D\}$ defines a Gabriel filter on R when is a star-operation on R. The corresponding torsion theory will be denoted by σ.

Proposition 3.5.22 *If R is σ-multiplication, then it is $*$-multiplication.*

Proof: Let $A = A^* \leq R$. Take $y \in Cl_\tau^R(A)$. Then, there exists $I \in \mathcal{F}$ such that $yI \subseteq A$. Since $I^* = D$, $(yI)^* \subseteq A^* = A$. By the properties of star-operations, $yI^* \subseteq A$ and hence, $yR \subseteq A$. It follows that $y \in A$ and A is σ-closed. By Theorem 3.5.18, there exists $B \in Frac(R)$ such that $Cl_\tau^R(B.A) = R$. Therefore, $B.A \in \mathcal{F}$ and consequently, $(B^*.A)^* = R$. It is deduced that A is $*$-invertible. ■

Given $A \in Frac(R)$, we denote by $A^{-1} = (R : A) = \{k \in K; Ak \subseteq R\}$. The mapping from $Frac(R)$ into itself which maps A into $A_v = R : (R : A)$ is called *v-operation* and it is a star-operation (see [51, Theorem 34.2]). Therefore, $R = R_v$. The fractional ideal A is called a *v-ideal* if $A = A_v$. It is said that R is v-multiplication if it is $*$-multiplication for the v-operation. Let $\chi_1(R)$ be the torsion theory cogenerated by the injective hull of K/R, $E(K/R)$. Then, an R-module is $\chi_1(R)$-torsion when $Hom_R(M, E(K/R)) = 0$. These two concepts are related in the next result. Before that, we recall some definitions.

Given a subring R of R', an element $a \in R'$ is *almost integral over* R if there exists a finitely generated submodule of the R-module R' containing all powers of a. The set R_0 of elements of R' which are almost integral over R is called the *complete integral closure* of R in R'. If $R = R_0$, it is said that R is *completely integrally closed in* R'. If R is completely integrally closed in its rings of quotients, it is said that R is *completely integrally closed.* A fractional ideal A of R is called *divisorial* if $A = (R : (R : A))$.

Theorem 3.5.23 *For an integral domain R the following sentences are equivalent*

1. *R is v-multiplication.*

2. *R is compeletely integrally closed.*

3. R is a $\chi_1(R)$-multiplication ring.

Proof: The first step is to prove that multiplication domains relative to $\chi_1(R)$ are just v-multiplication domains. Firstly, we will prove that $\mathcal{F}(\chi_1(R)) = \{A \leq R; A_v = R\}$. Suppose that $A \leq R$ verifies $A_v = R$. Let $J \leq R$ and $A \subseteq J$. Let $f : J/A \to K/R$ be a homomorphism of R-modules. By injectivity, we extend it to a morphism $f' : R/A \to E(K/R)$ such that $Im\ f = f'(J/A)$. Then, $A.Im\ f' = 0$ and hence, $Im\ f' \cap (K/R) \subseteq (R : A)/R$. Thus

$$
\begin{aligned}
Im\ f' \cap (K/R) &= R(Im\ f' \cap (K/R)) \\
&= A_v(Im\ f' \cap (K/R)) \subseteq (A_v.(R : A) + R)/R = 0.
\end{aligned}
$$

Because K/R is essential in its injective hull, $Im\ f' = 0$ and therefore, $f = 0$. This proves that $Hom_R(J/A, (K/R)) = 0$. It follows that $A \in \mathcal{F}(\chi_1(R))$. On the other hand, if $A \in \mathcal{F}(\chi_1(R))$, it is clear that $R = Cl^K_{\chi_1(R)}(A)$ and consequently, $A_v = R$.

Now, notice that $A \in Frac(R)$ is $\chi_1(R)$-invertible if and only if there is a fractional ideal B such that $Cl^K_{\chi_1(R)}(B.A) = R$, but it is equivalent to the existence of $B \in Frac(R)$ verifying $(B.A)_v = R$, that is, A is v-invertible. By [11, Proposition 3.2], for every v-invertible ideal A, it holds $A_v = Cl^K_{\chi_1(R)}(A)$. Now, it suffices to apply Proposition 3.5.21.

By [11, Theorem 3.1 (f)], R is completely integrally closed (c.i.c.) if and only if every fractional ideal is v-invertible. It suffices that v-ideals are v-invertible. By the first step, this is equivalent to every fractional ideal in R is $\chi_1(R)$-invertible. ■

From [69, Theorem 8.12], the following result is immediate.

Corollary 3.5.24 *$\chi_1(R)$-noetherian domains are exactly Krull domains.*

Corollary 3.5.25 *If the integral domain R is completely integrally closed, then its localizations in the height one primes are multiplication rings.*

Proof: By [44, Corollary 3.11], R_P is a discrete valuation domain for every height one prime ideal (divisorial prime ideal) of R and, therefore, in this case, R_P is multiplication. ■

Corollary 3.5.25 induces us to give the following definition.

Definition 3.5.26 *A ring R is said to be a divisorially multiplication ring if for every height one prime P of R, R_P is a multiplication ring.*

Recall that an *almost Dedekind domain* is a domain R verifying that R_P is a Dedekind domain for every prime ideal P.

Proposition 3.5.27 *If τ is a semicentered torsion theory on a τ-noetherian almost Dedekind domain R, then R is τ-multiplication.*

Proof: It is straightforward from Proposition 3.2.7. ■

Given any ring R and any hereditary torsion theory τ on R, there is another hereditary torsion theory τ_x on the polynomial ring $R[X]$ whose Gabriel filter is

$$\mathcal{F}(\tau_x) = \{J \leq R[X]; (J : f) \cap R \in \mathcal{F}(\tau)\ \forall f \in R[X]\}.$$

The next result provides a tool to obtain new relative multiplication rings.

Proposition 3.5.28 *The polynomial ring $R[X]$ is τ_x-multiplication if and only if the ring R is τ-multiplication.*

Proof: In the first step, we prove that for every ideal A of R, we have $Cl_{\tau_x}^{R[X]}(A[X]) = Cl_{\tau}^{R}(A)[X]$.

Let $f = r_0 + r_1x + r_2x^2 + \cdots + r_nx^n \in Cl_{\tau_x}^{R[X]}(A[X])$. Then, there exists $J \in \mathcal{F}(\tau_x)$ such that $fJ \subseteq A[X]$. It is deduced that every polynomial in fJ have coefficients in A. It follows that $r_i(J \cap R) \in A$ $(1 \leq i \leq n)$. As $J \cap R = (J : 1) \cap R \in \mathcal{F}(\tau)$, it holds that $r_i \in Cl_{\tau}^{R}(A)$ $(1 \leq i \leq n)$. Therefore, $f \in Cl_{\tau}^{R}(A)[X]$ and hence, $Cl_{\tau_x}^{R[X]}(A[X]) \subseteq Cl_{\tau}^{R}(A)[X]$.

Mow, take $f = r_0 + r_1x + \cdots + r_nx^n \in Cl_{\tau}^{R}(A)[X]$. Then, $r_i \in Cl_{\tau}^{R}(A)$ $(1 \leq i \leq n)$. Thus there exists $H \in \mathcal{F}(\tau)$ such that $r_iH \subseteq A$ $(1 \leq i \leq n)$. It follows that $f.H[X] \subseteq A[X]$. Moreover, given $g \in R[X]$, $(H[X] : g) \cap R = \{r \in R; rg \in H[X]\} \in \mathcal{F}(\tau)$ as it contains H. Therefore, $H[X] \in \mathcal{F}(\tau_x)$. Consequently, $f \in Cl_{\tau_x}^{R[X]}(A[X])$ and the equality holds.

Now, we suppose that $R[X]$ is τ_x-multiplication. Let A, B be τ_x-closed ideals of R with $A \subseteq B$. By the first step, $A[X]$ and $B[X]$ are τ-closed. It is clear that $A[X] \subseteq B[X]$. By hypothesis,

$$\begin{aligned} A[X] &= Cl_{\tau_x}^{R[X]}((A[X] : B[X]).B[X]) = Cl_{\tau_x}^{R[X]}((A : B)[X]).B[X]) \\ &= Cl_{\tau_x}^{R[X]}((A : B).B[X])) = Cl_{\tau}^{R}((A : B).B)[X]. \end{aligned}$$

Then, $A = Cl_{\tau}^{R}((A : B).B)$ and hence, R is τ-multiplication. This reasoning can be repeated to prove the other inclusion because of the first step. ■

Chapter 4

Multiplication objects in gr-R

Graded modules and rings appear in a natural way in commutative algebra, algebraic geometry, group representation, etc and they have been recently studied by many authors (cf. [29], [91], [112], [113] and their references). In [113], ideal arithmetical theory is applied in order to study the graded concepts of Dedekind domains and principal ideal domains. In [29], concepts related to that of multiplication ring such as graded Prüfer domain are investigated. Therefore, in this chapter, we study the graded version of the notions of multiplication module and ring. First, the basic definitions, the notation and the main facts on graded modules and rings, apart from several general properties, are exposed. In the second section, we shall research into local properties by localizing in graded prime ideals. The third section is devoted to the study of the graded concept corresponding to distributive module and ring, relating them to those of multiplication graded module and ring. Section 4.4 is concerning with multiplication graded rings, which are locally characterized, in the noetherian graded case. Section 4.5 deals with the question of when a graded ring is multiplication. It is solved when the grading group is abelian and torsionfree and for commutative twisted group algebras when the grading group is finite.

4.1 Properties of multiplication graded modules

After giving basic notation and definitions on graded modules and rings, we shall use that, in the strongly graded case, definitions of gr-multiplication module (multiplication graded) and of multiplication object in gr-R. We shall characterize gr-multiplication domains and some examples of gr-multiplication rings which are not multiplication will be studied.

Recall from [91] the basic definitions on graded rings and modules. A *ring R graded by a group G* will be denoted by $R = \oplus_{g\in G} R_g$, where R_g is an additive subgroup of R and $R_g.R_h \subseteq R_{gh}$ for every g, h in G. If the inclusion is an equality, then the ring is called *strongly graded.* If an element of R belongs to $\cup_{g\in G} R_g = h(R)$, then it is called *homogeneous.* An R-module M is said to be a *graded module* if $M = \oplus_{g\in G} M_g$ for a family of subgroups $\{M_g\}_{g\in G}$ of M such that $M_h.R_g \subseteq M_{hg}$ for every g, h in G. In a similar way, we define a *strongly graded module.* The ring R is strongly graded if and only if every graded R-module is strongly graded (cf. [91, Theorem A.I.3.4]). A *graded submodule* N of M ($N \leq_{gr} M$) is a submodule verifying $N = \oplus_{g\in G}(N \cap M_g)$. We write $\mathcal{L}^{gr}(M)$ for the lattice of all graded submodules of M. The identity element of G is denoted by e. R represents a commutative ring with an identity element. If M and N are two R-modules graded by an abelian group G, a morphism of graded R-modules $f : M \to N$ is a morphism of R-modules verifying $f(M_g) \subseteq N_g$ for every $g \in G$. The set of all these morphisms is denoted by $Hom_{gr-R}(M, N)$. Given a graded R-module M, the R-module $M(g)$, graded by means of $(M(g))_h = M_{gh}$ is called the *g-suspension* of M. The category of right R-modules graded by a group G and degree preserving homomorphisms is denoted by gr-R.

If M is an R-module and N is its submodule, the ideal $\{r \in R; M.r \subseteq N\}$ will be denoted by $(N : M)$. An ideal is *gr-maximal* if it is maximal in the lattice of graded ideals. The *graded Jacobson radical* of a ring R is defined as the intersection of all gr-maximal ideals. It will be denoted by $J^{gr}(R)$. A graded ring having an only gr-maximal ideal is called *graded local* (*gr-local*). If P is a graded ideal of a graded ring R, P is called *graded prime* if it verifies that $I.J \subseteq P$ implies that $I \subseteq P$ or $J \subseteq P$ for any graded ideals I, J of R. The set of all graded prime ideals is denoted by $Spec^{gr}(R)$. The set of gr-maximal ideals will be denoted by $Max^{gr}(R)$. Recall that a graded ring is *gr-noetherian* if every graded ideal is finitely generated.

Given a ring R graded by G and a multiplicative subset $S \subseteq h(R)$, the ring of fractions $S^{-1}R$ turns into a ring graded by G by means of $(S^{-1}R)_g = \{\frac{r}{s}; s \in S, r \in R, g = (deg\ s)^{-1} deg\ r\}$ for every $g \in G$, where $deg\ s$ represents the degree of s. Moreover, in the strongly graded case, $S^{-1}R$ is also strongly graded (see [91, Proposition A.I.6.2 and Corollary A.I.6.3]). When $S = h(R) - P$ for some graded prime ideal P of R, we will write R_P^g for $S^{-1}R$. Recall that $S^{-1}M$ can be defined as $S^{-1}R \otimes_R M$, when M is an R-module. Throughout this chapter, unless stated otherwise, every graded ring is non-trivially graded.

Lemma 4.1.1 *If M is a graded R-module and N is a submodule of M, then $(N : M)$ is a graded ideal.*

Proof: Take a homogeneous element of M, say m. Let $x \in (N : M)$. Since $x \in R$, $x = x_{g_1} + \cdots + x_{g_n}$ where $x_{g_i} \in R_{g_i}$ $(1 \leq i \leq n)$. Then, $m.x \in N$. Therefore, in the graded module M/N, $(m.x_{g_1} + N) + \cdots + (m.x_{g_n} + N) = N$. By taking into account that this sum is direct, every $m.x_{g_i} \in N$. Thus $x_{g_i} \in (N : M) \cap R_{g_i}$. As a consequence, $(N : M) \subseteq \oplus_{g \in G}(N : M) \cap R_g$ and the other inclusion is trivial. Therefore, $(N : M)$ is a graded ideal. ■

Definition 4.1.2 *A graded R-module M is said to be a multiplication graded module (gr-multiplication) if for every graded submodule N, there exists an ideal I of R such that $N = M.I$.*

Remark. Obviously, every multiplication module which is graded is a gr-multiplication module. Moreover, if $N = M.I$ for some ideal I of R, then $N = M.(N : M)$ and, therefore, the ideal I of the definition can be taken graded (cf. Lemma 4.1.1). We will show that there are examples of multiplication graded modules that are not multiplication modules.

Remark. If R is commutative, gr-R is a commutative Grothendieck category when R is strongly graded, as it is well-known that, in this case, R is a generator of gr-R. However, in [78, Example 3.5], it is possible to check that gr-R can be a commutative Grothendieck category without R is strongly graded. The example consists of considering the infinite product, B, of copies of any commutative ring, A, and the graded ring $R \times \{0\}$ of kind $G = \mathbb{Z}_2$.

Then, $U = R(0) \oplus R(1)$ (cf. [78, p. 289]), is a generator, with endomorphism ring isomorphic to A, and however, R is not strongly graded. Nevertheless, the category gr-R, when R is commutative is always monoidal.

The following Theorem provides us examples of gr-multiplication modules.

Theorem 4.1.3 *Let R be a strongly graded ring and let M be a graded R-module. Then, M is a gr-multiplication module if and only if M_e is multiplication as an R_e-module.*

Proof: By [91, Proposition 1.3.4], we have an equivalence of categories between gr-R and Mod-R_e, given by the functors tensor product $- \otimes_{R_e} R$ and restriction to the degree e component, $(-)_e$. By Proposition 2.2.7, M is a multiplication object in gr-R if and only if M_e is a multiplication object in Mod-R_e. Moreover, this equivalence is monoidal. By Proposition 2.1.6, the result follows. ■

When working with graded modules, there are three related categories, namely, gr-R, Mod-R_e and Mod-R. In this way, we have the next result.

Proposition 4.1.4 *If $M \in$ gr-R and M is multiplication as an R_e-module, then M is a multiplication R-module.*

Proof: Let N be an R-submodule of M. Since $N.R_e = N$ and by hypothesis, $N = M.(N :_{R_e} M) \subseteq M.(N :_R M) \subseteq N$. Therefore, M is multiplication as an R-module. ■

If $M \in$ gr-R and N is a graded submodule of M, then M/N can be graded by means of $(M/N)_g = (M_g + N)/N$ for every $g \in G$. We give a proof where graded techniques can be appreciated.

Proposition 4.1.5 *If M is a gr-multiplication module and N is a graded submodule of M, then M/N es gr-multiplication.*

Proof: Let L/N be a graded submodule of M/N. Then, $L/N = \oplus_{g\in G}((L/N)\cap(M/N)_g) = \oplus_{g\in G}((L/N)\cap(M_g+N)/N) = \oplus_{g\in G}((L\cap M_g)+N)/N$. Thus $L = \sum_{g\in G}((L\cap M_g)+N) = (\sum_{g\in G}(L\cap M_g))+N = (\oplus_{g\in G}(L\cap M_g))+N$. Since N is graded, $N = \oplus_{g\in G}N_g \subseteq \oplus_{g\in G}(L\cap M_g)$. Therefore, $L = \oplus_{g\in G}(L\cap M_g)$ is also graded. By hypothesis, there exists an ideal A of R such that $L = M.A$ and hence, $L/N = (MA)/N = (M/N)A$. ■

4.2 Graded localization

We shall use a graded version (with gr-multiplication modules) of Nakayama's Lemma as a technical tool to prove properties of local or global kind. Some characterization theorems similar to those of absolute case are obtained (see, for instance, [31]). Finitely generated gr-multiplication modules will be locally characterize in several different ways. The section is completed with some illustrative examples.

Proposition 4.2.1 *If M is a gr-multiplication R-module and $A \subseteq J^{gr}(R)$, then $M = MA$ implies that $M = 0$.*

Proof: Let $x \in h(M)$. Then, the ideal xR is graded. Since M is gr-multiplication, there exists an ideal B of R such that $xR = MB$. It holds $xR = MAB = MBA = xRA = xA$ and therefore, there exists $a \in R$ such that $x = xa$. Since x is homogeneous, $a \in R_e \cap J^{gr}(R)$. Hence, $x(1-a) = 0$. By [91, Lemma A.I.7.4 (6)], $1-a$ is invertible because $(1-a)a \in J^{gr}(R)$. It follows that $x = 0$. Therefore, $h(M) = 0$ and since M is graded, $M = 0$. ■

Proposition 4.2.2 *If M is a gr-multiplication R-module and $S \subseteq h(R)$ is a multiplicative subset of R, then $S^{-1}M$ is gr-multiplication as an $S^{-1}R$-module.*

Proof: Take a graded $S^{-1}R$-submodule of $S^{-1}M$. It has to be of the form $S^{-1}N$ with N a graded submodule of M (by the flatness of $S^{-1}R$). By hypothesis, there exists $A \leq R$ such that $N = MA$. By applying S^{-1}, we have $S^{-1}N = S^{-1}(MA) = S^{-1}MS^{-1}A$, where $S^{-1}A \leq S^{-1}R$ and the proof ends. ■

Assume that M is a graded R-submodule. Given a graded prime ideal P, i. e., $P \in Spec^{gr}(R)$, we consider the set

$$T_P(M) = \{m \in M; \exists c \in h(R) - P \ \wedge \ mc = 0\}.$$

It is easy to check that $T_P(M)$ is a graded R-submodule of M. If $M = T_P(M)$, we will say that M is *graded P-torsion.* If there exist $x \in h(M)$ and $c \in h(R) - P$ such that $Mc \subseteq xR$, we will say that M is *graded P-cyclic.* The following theorem is a graded version of [31, Theorem 1.2].

Theorem 4.2.3 *If M is a graded R-module, then M is gr-multiplication if and only if for every graded prime P, M is graded P-torsion or graded P-cyclic.*

Proof: Suppose that M is gr-multiplication and let $P \in Spec^{gr}(R)$. Firstly, we consider the case in which $MP = M$. Given $m \in h(M)$, the submodule mR is graded. Since M is gr-multiplication, there exists $A \leq R$ such that $mR = MA = MPA = MAP = mP$. Hence, there exists $c \in h(R) - P$ verifying that $mc = 0$. If m is any element of M, the above property is also true. It suffices to multiply the different elements of $h(R) - P$ vanishing its homogeneous components. Now, consider that $MP \neq M$. There exists $x \in h(M) - MP$ and, since xR is graded, $xR = MB$, where B is an ideal that is possible to choose graded (for example, $(xR : M)$). If $B \subseteq P$, then $x \in MP$, a contradiction. Therefore, $B \not\subseteq P$. As a consequence, there exists $c \in h(B) - P$ and it holds $Mc \subseteq MB = xR$.

We will prove the converse implication. Let N be a graded submodule of M. Let $A = (N : M)$. Let $n \in h(N)$ and $K = (MA : nR)$, which, therefore, is a graded ideal. Suppose that $K \neq R$. Then, there will be exist a graded maximal P containing K. If M is graded P-torsion, then there exists $c \in h(R) - P$ such that $nc = 0$ and it follows that $c \in K - P$, a contradiction. By hypothesis, there exist $x \in h(M)$ and $c \in h(R) - P$ such that $Mc \subseteq xR$. Thus, Nc is a graded submodule of xR which is gr-multiplication because x is homogeneous and cyclic. Hence, there exists $J = (Nc : xR)$ such that $Nc = xJ$. It holds that $MJc = McJ \subseteq xJ \subseteq N$. It follows that $Jc \subseteq A$. The element nc^2 belongs to $Nc^2 = xcJ \subseteq MA$. As a result, $c^2 \in K \subseteq P$, which is another contradiction. It follows that $K = R$ and it is straightforward that $R = (MA : N)$ and, therefore, $N \subseteq M(N : M)$. Since the other inclusion is always true, the proof ends. ■

Corollary 4.2.4 *A graded R-module M is gr-multiplication if and only if for every $P \in Spec^{gr}(R)$, M is graded P-torsion or there exist $N \leq_{gr} M$ and $c \in h(R) - P$ such that $Mc \subseteq N$.*

Proof: The necessity of the condition in order to obtain that M is gr-multiplication is obvious by Theorem 4.2.3. To prove the sufficiency, take

$P \in Spec^{gr}(R)$ and suppose that M is not graded P-torsion. Then, by hypothesis, there exists $c \in h(R) - P$ such that $Mc \subseteq L$, L being a gr-multiplication submodule of M. Since M is not graded P-torsion, L is not graded P-torsion. By Theorem 4.2.3, there exist $x \in h(L) - P$ and $d \in R - P$ such that $Ld \subseteq xR$. Thus $Mcd \subseteq xR$ and M is graded P-cyclic and therefore, gr-multiplication. ■

Corollary 4.2.5 *If the graded R-module* $M = \sum_{k \in K} m_k R$, *K being any set of indices and* m_k $(k \in K)$ *a set of homogeneous elements of* M, *then* M *is gr-multiplication if and only if for every* $h \in K$ *exists* $I_k \leq R$ *such that* $m_k R = MI_k$.

Proof: If M is gr-multiplication, then the stated condition is true because every $m_k R$ is a graded submodule. Suppose that the aforementioned condition is true and takes $P \in Spec^{gr}(R)$. If some $I_k \not\subseteq P$, then $m_k R = MI_k$ and in this case, since there exists $c \in h(I_k) - P$, we have that M is graded P-cyclic. If not, $I_k \subseteq P$ for every $k \in K$ and hence, $M = MP$. Then, $m_k R = MI_k = MPI_k = m_k P$. It follows that $m_k \in T_P(M)$ and therefore, $M = T_P(M)$. By Theorem 4.2.3, M is gr-multiplication. ■

Corollary 4.2.6 *If* M *is a gr-multiplication module and the ideal* I *is gr-multiplication, then the R-module* MI *is gr-multiplication.*

Proof: Let $P \in Spec^{gr}(R)$. If $I = T_P(I)$, then we have $T_P(MI) = MI$. For $M = T_P(M)$, it happens the same thing. Suppose that $I \neq T_P(I)$ y $M \neq T_P(M)$. By Theorem 4.2.3, there exist $c_1, c_2 \in h(R) - P$, $a \in h(I)$ and $x \in h(M)$ such that $Ic_1 \subseteq aR$ and $Mc_2 \subseteq xR$. Therefore, $Mc_1c_2 \subseteq Mac_2 \subseteq$

xaR. Moreover, $xa \in h(MI)$. It follows that MI is graded P-cyclic and it suffices to apply Theorem 4.2.3 again. ■

By using Theorem 4.2.3, we can prove the following characterization of gr-multiplication module in which $ann(M)$ is the *annihilator* of M, i.e., $ann(M) = \{r \in R; Mr = 0\}$.

Theorem 4.2.7 *If M is a graded R-module with $ann(M) = 0$, then M is gr-multiplication if and only if:*

1. *$\cap_{k\in K}(MI_k) = M(\cap_{k\in K} I_k)$ for every family I_k $(k \in K)$ of graded ideals of R.*
2. *for every graded submodule N of M and every graded ideal A of R such that $N \subset MA$, there exists a grade ideal $B \subset A$ verifying $N \subseteq MB$.*

Proof: Suppose the right side of the equivalence is true. Let $N \leq_{gr} M$ and we call $C = \{I \leq_{gr} R; N \subseteq MI\}$. It is clear that $C \neq \emptyset$ as $R \in C$. Let I_k $(k \in K)$ be a non-empty family of elements of C. By condition 1, the ideal $\cap_{k\in K} I_k$, which is graded, is inside C. By applying Zorn's Lemma, there exists some minimal element of C, say A. Thus $N \subseteq MA$. If $N \neq MA$, by condition 2, there exists $B \leq_{gr} R$ such that $N \subseteq MB$. Then, $B \in C$, which contradicts the choice of A. Consequently, $N = MA$ and M is a gr-multiplication module.

Now, suppose that M is gr-multiplication. Let I_k $(k \in K)$ a collection of graded ideals of R. We call $I = \cap_{k\in K} I_k$. The inclusion $MI \subseteq \cap_{k\in K}(MI_k)$ always holds. Since MI_k is a grade module, we can consider $x \in h(\cap_{k\in K}(MI_k))$. Let $H = (MI : xR)$ which will be graded. Assume that $K \neq R$. Then, there exists $P \in Spec^{gr}(R)$ containing H. Obviously, $x \notin T_P(M)$, as otherwise, we would find an element in $H - P$. By Theorem 4.2.3, M is graded P-cyclic. Thus there exist $m_0 \in h(M)$ and $c \in h(R) - P$ verifying $Mc \subseteq m_0R$. Then, $x \in \cap_{k\in K} m_0 I_k$. It follows that for every $k \in K$, there exists $a_k \in I_k$ such that $xc = m_0a_k$. We choose $k_0 \in K$. Then, $m_0a_{k_0} = m_0a_k$ for every $k \in K$. Therefore, $m_0(a_{k_0} - a_k) = 0$ for every $k \in K$. We have $M(a_{k_0} - a_k)c = Mc(a_{k_0} - a_k) \subseteq m_0R(a_{k_0} - a_k) = 0$. Since $ann(M) = 0$, $c(a_{k_0} - a_k) = 0$. Hence, $ca_{k_0} = ca_k \in I_k$ for every $k \in K$. As a consequence, $ca_{k_0} \in I$. It follows that $xc^2 = m_0a_{k_0}c \in MI$. Therefore, $c^2 \in K \subseteq P$, a contradiction. As a result, $H = R$. It is easily deduced that $x \in MI$ and hence, the condition 1 is true. Let $N \leq_{gr} M$ and $A \leq_{gr} R$ such that $N \subset MA$. By hypothesis, there exists a graded ideal C verifying $N = MC$. We take $B = A \cap C$ and since $B \subset A$, we have $N = MA \cap MC = M(A \cap C) = MB$ by condition 1, which proves condition 2. ■

If we consider the ring $R' = R/(ann(M))$, R' is a graded ring, according to Proposition 4.1.5. It is easy to check that M is gr-multiplication R-module if and only if M is gr-multiplication as an R'-module. Since the sum of graded ideals is another graded ideal, the next result is evident.

Corollary 4.2.8 *An R-module M is un gr-multiplication if and only if*

1. $\cap_{k\in K}(MI_k) = M(\cap_{k\in K}[I_k + ann(M)])$ *for every non-empty collection og graded ideals.*

2. *para for every graded submodule N of M and graded ideal A of R such that $N \subset MA$, there exists a graded ideal B with $B \subset A$ and $N \subseteq MB$.*

The following examples, inspired in [31], show that the two conditions of the Theorem 4.2.7 are required. Let $R = \frac{K[x,y]}{(x^2,y^2,xy)}$ and let $P = \frac{(x,y)}{(x^2,y^2,xy)}$ which is maximal in R. We call $R' = R^g_P$ and M is its unique maximal ideal. We consider in R, R' and in the R'-module M the trivial graduation. We can take a non-zero element $m \in M$. Then, mR cannot be equal to MI for any ideal I of R as $MI \subseteq M^2 = 0$ if $I \neq R$ and $M \neq mR$ because M is not cyclic. Therefore, M is not (gr-)multiplication. However, the first condition of the theorem is true by using that $M^2 = 0$. Now, consider the ring R of matrices of the form

$$\begin{pmatrix} z & q \\ 0 & z \end{pmatrix}$$

where $z \in \mathbb{Z}$ y $q \in \mathbb{Q}$. Let M be the ideal of all matrices of R with $z = 0$. We consider the trivial graduation in R and in M. Then, M is not gr-multiplication as if N is the submodule of M with elements of the form

$$\begin{pmatrix} 0 & z \\ 0 & 0 \end{pmatrix}$$

with $z \in \mathbb{Z}$. Ideals of R, as it can check it by following [56, Proposition 4.1], have their elements of the form

$$\begin{pmatrix} c & q \\ 0 & c \end{pmatrix}$$

with $c \in C \leq \mathbb{Z}$ and $q \in \mathbb{Q}$ or of the form

$$\begin{pmatrix} 0 & l \\ 0 & 0 \end{pmatrix}$$

with $l \in L_{\mathbb{Z}} \leq \mathbb{Q}_{\mathbb{Z}}$. Therefore, supposing that there exists $A \leq R$ such that $N = MA$ implies $\mathbb{Z} = 0$ or $\mathbb{Z} = \mathbb{Q}$ and both options are absurd. However,

M verifies the second condition of the theorem. In fact, if $N \leq_{gr} M$, its elements are of the form

$$\begin{pmatrix} 0 & l \\ 0 & 0 \end{pmatrix}$$

with $l \in L$, which is a $\mathbb{Z}$-submodule of $\mathbb{Q}$. If $N \subset MA$ for some ideal A of R, the elements of A must be of the second kind and moreover, $L \subset \mathbb{Q}$. Then, the ideal B with elements of the form

$$\begin{pmatrix} c & 0 \\ 0 & c \end{pmatrix}$$

with $c \in C$ verifies the conditions of the ideal B of condition 2 of the theorem.

The concepts of finitely cogenerated module and ring appear related to those of multiplication module and ring in [32, Corollary 1.8]: It is possible to achieve a graded version.

Definition 4.2.9 *A graded R-module M is called graded finitely cogenerated if for every non-empty family of graded submodules N_k $(k \in K)$ such that $\cap_{k \in K} N_k = 0$, there exists a finite subset F of K verifying $\cap_{k \in F} N_k = 0$.*

Definition 4.2.10 *The graded ring R is graded finitely cogenerated if it is graded finitely cogenerated as an R-module.*

Corollary 4.2.11 *Let M be a gr-multiplication R-module verifying ann(M) = 0. Then, M is gr-finitely cogenerated if and only if R is gr-finitely cogenerated.*

Proof: Suppose that M is gr-finitely cogenerated. Let $I_{k\in K}$ $(k \in K)$ a family of graded ideals of R such that $\cap_{k\in K} I_k = 0$. By Theorem 4.2.3, $\cap_{k\in K}(MI_k) = 0$. Since M is gr-finitely cogenerated, there exists a finite subset $F \subseteq K$ such that $\cap_{k\in F}(MI_k) = 0$. By applying the same theorem again, $M(\cap_{k\in F} I_k) = 0$ and since $ann(M) = 0$, $\cap_{k\in F} I_k = 0$. Therefore, R is gr-finitely cogenerated.

Conversely, suppose that R is gr-finitely cogenerated. Let $N_{k\in K}$ $(k \in K)$ a non-empty family of graded submodules of M such that $\cap_{k\in K} N_k = 0$. For every $k \in K$, there exists a graded ideal A_k of R such that $N_k = MA_k$. Then, it is clear that $M(\cap_{k\in K} A_k) \subseteq \cap_{k\in K} N_k$. By hypothesis, $\cap_{k\in F} A_k = 0$. By Theorem 4.2.7, $\cap_{k\in F} N_k = \cap_{k\in F}(MA_k) = M(\cap_{k\in F} A_k) = 0$. As a consequence, M is gr-finitely cogenerated. ■

Lemma 4.2.12 *If M is a graded R-module, $M = 0$ if and only if $M_P^g = 0$ for every $P \in Spec^{gr}(R)$.*

Proof: If $M = 0$, then $M_P^g = 0$ for every $P \in Spec^{gr}(R)$ obviously. We prove the converse implication. Suppose that there exists a non-zero element of $x \in M$. Consider the ideal $ann(x)$, which is not R because $1 \notin ann(x)$. Therefore, there exists $P \in Spec^{gr}(R)$ such that $x \in P$. Since $\frac{x}{1} \in M_P^g = 0$, it follows that there exists $s \in h(R) - P$ such that $xs = 0$. Thus $s \in (h(R) - P) \cap ann(x)$, a contradiction. ■

Given a morphism $\Phi : M \to N$ of graded R-modules and $P \in Spec^{gr}(R)$, we denote by $\Phi_P : M_P^g \to N_P$ the morphism of graded R_P^g-modules defined by $\Phi(\frac{m}{s}) = \frac{\Phi(m)}{s}$ for every $m \in M$ and $s \in h(R) - P$.

Lemma 4.2.13 *If $\Phi : M \to N$ is a morphism of graded R-modules, then Φ is bijective if and only if so is Φ_P for every $P \in Spec^{gr}(R)$.*

Proof: Since $Ker\ \Phi_P = (Ker\ \Phi)_P$ for every $P \in Spec^{gr}(R)$, it is immediate that Φ is injective if and only if the Φ_P are injective.

Analogously, it is proved that the surjectiveness, by taking into account that $(\Phi(M))_P = \Phi_P(M)$. ∎

Lemma 4.2.14 *If M is a graded R-module and N is a graded submodule of M, then $M = N$ if and only if $N_P = M_P^g$ for every $P \in Spec^{gr}(R)$.*

Proof: Consider the inclusion i from N into M, which is a morphism of graded R-modules. $M = N$ if and only if i is bijective. By Lemma 4.2.13, that is true if and only if i_P, the inclusion from N_P into M_P^g is bijective for every $P \in Spec^{gr}(R)$, which is equivalent to say that $M_P^g = N_P$ for every prime ideal $P \in Spec^{gr}(R)$. ∎

Proposition 4.2.15 *If M is a finitely generated module, then M is gr-multiplication as an R-module if and only if M_P^g is gr-multiplication as an R_P^g-module for every $P \in Spec^{gr}(R)$.*

Proof: Suppose that M is gr-multiplication. By Proposition 4.2.2, for $S = h(R) - P$, it is obtained that M_P^g's are gr-multiplication R_P^g-modules. To prove the converse implication, we take a graded submodule N of M. It holds that $N_P = M_P^g(N_P :_{R_P^g} M_P^g)M_P^g$ for every graded prime ideal P. Since M is finitely generated, $N_P = [M(N :_R M)]_P$. By Lemma 4.2.14, $N = M(N :_R M)$ and therefore, M is gr-multiplication. ∎

Definition 4.2.16 *An R-module M is said to be graded cyclic (gr-cyclic) if it can be expressed as mR for some $m \in h(M)$.*

It is clear that every gr-cyclic R-module is cyclic. However, the converse is not true. Consider the polynomial ring with integer coefficients, $R = \mathbb{Z}[X]$, graded in the usual form by $\mathbb{Z}$. The ideal $(1+X)R$ is cyclic but not gr-cyclic.

Proposition 4.2.17 *A graded finitely generated R-module M is gr-multiplication if and only if M_P^g is a gr-cyclic R_P^g-module for every $P \in Spec^{gr}(R)$.*

Proof: The sufficiency is immediately deduced from Proposition 4.2.15 because every cyclic module is multiplication and, hence, it is gr-multiplication. We will prove the necessity. Let $P \in Spec^{gr}(R)$. If $M = 0$, then it is clear. If $M \neq 0$, by Proposition 4.2.1, there exists $x \in h(M_P^g) - MP$ such that $xR_P^g = M_P^g A$ for some $A \leq R_P^g$. However, $A \not\subseteq PR_P^g$ because then $x \in M_P^g P$. Since A can be chosen graded, it follows that $A = R$ and it holds that $M_P^g = xR_P^g$. ■

4.3 Distributive graded modules

We shall locally characterize, by means of localization in the graded prime ideals, τ-distributive modules and, from Proposition 4.2.17, we shall prove the relationship between distributive graded modules (gr-distributive) and gr-multiplication modules.

Recall that a graded R-module M is *gr-distributive* if the lattice $\mathcal{L}^{gr}(M)$ of its graded submodules is distributive, i. e., it has distributive property of the sum with respect to the intersection (or, equivalently, of the intersection with respect to the sum). The following Lemma is a graded version of [106, Lemma 1.4].

Lemma 4.3.1 *A graded R-module M is gr-distributive if and only if for every graded module N and every $f \in Hom_{gr-R}(N, M)$, it holds $f^{-1}(A + B) = f^{-1}(A) + f^{-1}(B)$ for every $A, B \in \mathcal{L}^{gr}(M)$.*

Proof: Suppose that M is gr-distributive. Let N be another graded module, $f \in Hom_{gr-R}(N, M)$ and $A, B \in \mathcal{L}_{gr}(M)$. Since $f(N) \in \mathcal{L}^{gr}(M)$, by applying the hypothesis, we obtain $(A + B) \cap f(N) = (A \cap f(N)) + (B + f(N))$. Therefore,

$$\begin{aligned} f^{-1}(A + B) &= f^{-1}((A + B) \cap f(N)) \\ &= f^{-1}((A \cap f(N)) + (B \cap f(N))) \\ &= f^{-1}(ff^{-1}(A) + ff^{-1}(B)) \\ &= f^{-1}ff^{-1}(A) + f^{-1}ff^{-1}(B) \\ &= f^{-1}(A) + f^{-1}(B). \end{aligned}$$

To prove the converse implication, we take $A, B, C \in \mathcal{L}^{gr}(M)$ and consider the inclusion $i : C \to M$. By hypothesis,

$$(A + B) \cap C = i^{-1}(A + B) = i^{-1}(A) + i^{-1}(B) = (A \cap B) + (B \cap C).$$

■

Lemma 4.3.2 *Given $M, N \in$ gr-R and $f, g \in Hom_{gr-R}(M, N)$, if N is gr-distributive, then $M = g^{-1}(Im\ f) + f^{-1}(Im\ g)$.*

Proof: Since $Im(f+g) \subseteq Im\ f + Im\ g$, we have

$$\begin{aligned} M &= (f+g)^{-1}(Im\ f + Im\ g) \\ &= (f+g)^{-1}(Im\ f) + (f+g)^{-1}(Im\ g) \\ &= g^{-1}(Im\ f) + f^{-1}(Im\ g) \end{aligned}$$

by Lemma 4.3.1. ■

Recall that a graded submodule L of M is *gr-inessential* when $L+N \neq M$ with $N \in \mathcal{L}^{gr}(M)$ and $N \neq M$. A graded ideal is gr-inessential if and only if it is contained in $J^{gr}(R)$.

Lemma 4.3.3 *Let $M, N \in$gr-R and $f, g \in Hom_{gr-R}(M, N)$. If every proper graded submodule of M is gr-inessential and N is gr-distributive, then $Im\ f \subseteq Im\ g$ or $Im\ g \subseteq Im\ f$.*

Proof: By Lemma 4.3.2, $M = g^{-1}(Im\ f) + f^{-1}(Im\ g)$. By hypothesis, $M = g^{-1}(Im\ f)$ or $M = f^{-1}(Im\ g)$. Thus $Im\ f \subseteq Im\ g$ or $Im\ g \subseteq f$. ■

Lemma 4.3.4 *If R is a gr-local ring and M is a graded R-module, then M is gr-distributive if and only if $\mathcal{L}^{gr}(M)$ is totally ordered under the inclusion.*

Proof: Suppose that M is gr-distributive. It suffices to take $M = R$ in Lemma 4.3.3 and to check that if the gr-cyclic ideals are ordered, then so is $\mathcal{L}^{gr}(R)$. The converse implication is also straightforward. ■

Lemma 4.3.5 *If R is a gr-local ring and M is a graded R-module such that every finitely generated graded submodule is gr-cyclic, then $\mathcal{L}^{gr}(M)$ is totally ordered under the inclusion.*

Proof: Suppose that there exist $L, N \in \mathcal{L}^{gr}(M)$ such that $L \not\subseteq N$ and $N \not\subseteq L$. Then there exist $l \in h(L) - N$ and $n \in h(N) - L$. Then $lR + nR$ is graded because it is a sum of graded ideals and obviously, it is finitely generated. By hypothesis, $lR + nR = xR$ with $x \in h(R)$. Then, there exist homogeneous elements of R, namely, p, q, r, s such that $l = xp$, $n = xq$ y $x = lr + ns$. By substituting, $x(1 - rp - sq) = 0$. Since these elements are all homogeneous, $1 - rp - sq \in R_e$. As it is not a unit, because $x \neq 0$, it will be inside the only gr-maximal ideal P. Therefore, $rp \notin P$ or $sq \notin P$. If the first possibility is true, it follows that p is a unit and then, $n \in xR = lR$, a contradiction. With the second possibility we come to the same conclusion as before. ■

Lemma 4.3.6 *A graded R-module M is gr-distributive if and only if the graded R^g_P-module M^g_P is gr-distributive for every $P \in Spec^{gr}(R)$.*

Proof: The implication from the left to the right side is true by Proposition 4.2.2. For the converse implication, given $L, N, K \in \mathcal{L}^{gr}(M)$, the inclusion $(L \cap N) + (L \cap K) \subseteq L \cap (N + K)$ is always true and both of them are graded submodules. The other inclusion follows from the hypothesis and Lemma 4.2.14. ■

Proposition 4.3.7 *The graded R-module M is gr-distributive if and only if for every finitely generated graded submodule N, the R^g_P-module N^g_P is gr-cyclic.*

Proof: Suppose that M is gr-distributive. It is clear, from the definition, that any graded submodule N of M is also gr-distributive. Proposition 4.2.2

implies that the R_P^g-module N_P^g is gr-distributive. By Lemma 4.3.4, the lattice of submodules of N_P^g for any $P \in Spec^{gr}(R)$ is totally ordered under the inclusion, in particular, N_P^g is gr-cyclic.

To prove the converse implication, given $P \in Spec^{gr}(R)$, by Lemma 4.3.5, the lattice of R_P^g-submodules of M_P^g is totally ordered under inclusion. By Lemma 4.3.4, M_P^g is gr-distributive as an R_P^g-module and by Lemma 4.3.6, the proof ends. ■

Proposition 4.3.8 *A graded R-module M is gr-distributive if and only if every finitely generated graded submodule is a gr-multiplication module.*

Proof: Suppose that M is gr-distributive. Then, every finitely generated graded submodule N verifies that N_P^g is gr-cyclic as an R_P^g-module by Proposition 4.3.7 and hence, it is multiplication and consequently, it is gr-multiplication as an R_P^g-module. By Proposition 4.2.15, N is gr-multiplication.

Conversely, by hypothesis and by Proposition 4.2.17, every finitely generated graded submodule N verifies that the R_P^g-module N_P^g is gr-cyclic (for every $P \in Spec^{gr}(R)$). By Proposition 4.3.7, M is gr-distributive. ■

4.4 Multiplication graded rings

Multiplication graded domains will be characterized. Several examples of multiplication graded rings (gr-multiplication) that are not multiplication will be exposed. We shall obtain local characterizations of multiplication graded rings in the graded noetherian case. Multiplication objects in the categories Mod-R_e, Mod-R and gr-R will be related. Noetherian rings graded by a torsionfree abelian group that are multiplication are determined. If the grading group is finite, multiplication rings are characterized in terms of R_e and we give other sufficient conditions in order to be multiplication a graded ring.

Definition 4.4.1 *A ring R is said to be a gr-multiplication ring if every graded ideal is gr-multiplication as an R-module.*

It is also possible to characterize multiplication graded domains in an analogous way to the ungraded case as we can see in the following proposition. Recall that a *graded Dedekind domain* (gr-Dedekind domain) is an integral domain in which every non-zero graded ideal is invertible (cf. [91] and its references).

Proposition 4.4.2 *gr-multiplication domains are just gr-Dedekind domains.*

Proof: Let $R = \oplus_{g\in G} R_g$ be a gr-Dedekind domain (not necessarily strongly graded). Then every graded ideal is invertible and, therefore, it is projective and hence, it is multiplication. As it has been expressed above, this implies that it is gr-multiplication.

Now, suppose that R is a gr-multiplication domain. Let A be a non-zero graded ideal of R. Since it is non-zero, there exists a homogeneous element $a_g \in A \cap R_g$. Obviously, $a_g.R \subseteq A$. By hypothesis, there exists a graded ideal C of R such that $a_g.R = A.C$. But $a_g.R$ is invertible (because R is a domain) and hence, so is A. ■

Examples. In [112], it is proved that if R is a Dedekind domain, then $R[\mathbb{Z}]$ is a gr-Dedekind domain. Therefore, $\mathbb{Z}[\mathbb{Z}]$ is a gr-Dedekind domain and hence, it is a gr-multiplication ring. However, it is not a multiplication domain because it is not a Dedekind domain (see [52, Theorem 13.8]). Thus there are gr-multiplication rings that are not multiplication rings.

Another kind of gr-multiplication rings are graded von Neumann regular rings or, more briefly, gr-regular rings. A ring R is *gr-regular* if for any homogeneous element x of R, there exists $y \in R$ such that $x = xyx$. By [91, Remarks (4), p. 258], every graded ideal is generated as an ideal by its part of degree e. Now, it is easy to prove that every graded ideal of a gr-regular ring (not necessarily strongly graded) is gr-multiplication. Moreover, by [91, Corollary C.I.5.3], if the ring is strongly graded, then it is gr-regular if and only if the component of degree e is regular. Therefore, the group ring $(\Pi_{i=1}^{\infty}\mathbb{Z}_2)[\mathbb{Z}]$ is gr-multiplication but not multiplication as [52, Theorem 18.10] establishes that group rings $R[G]$ with G torsion-free which are multiplication are those in which $G = \mathbb{Z}$ and R is a finite product of fields.

Now, we will characterize the gr-noetherian gr-multiplication by localization. Before that, we recall the graded version of discrete valuation domain (gr-DVD). If K is a graded field, a graded subring of K, R, is a *discrete graded-valuation domain* (gr-DVD, see [91, p. 164]) if for any homogeneous element x of K, it holds that x or x^{-1} belongs to R, $\mathbb{Z}$ being its valuation group. We define the graded version of a special primary ring. A graded ring is said to be a *graded special primary ring* (gr-SPIR) if it is gr-local and every proper graded ideal is a power of its only gr-maximal ideal. Then, it holds the following easy consequence of Theorem 4.1.3.

Proposition 4.4.3 *If R is a gr-noetherian, then R is gr-multiplication if and only if R^g_P is a gr-multiplication ring for every $P \in Spec^{gr}(R)$.*

Proof: It is obvious that if R is gr-noetherian, then the R^g_P's are gr-noetherian as well as ideals of R^g_P are localized from ideals of R. R is gr-multiplication when every graded ideal I is gr-multiplication and by Proposition 4.2.15, this happens if and only if I_P is gr-multiplication as an R^g_P-module for every graded prime ideal P. Since every ideal of R^g_P is the localized one of an ideal of R, it follows that the above sentence is equivalent to say that every R^g_P is a gr-multiplication ring. ∎

Lemma 4.4.4 *If R is a gr-multiplication ring and P is any gr-maximal ideal of R, then between P^{n+1} and P^n there is no graded ideal strictly contained for every positive integer n.*

Proof: Suppose that A is a graded ideal such that $P^{n+1} \subseteq A \subseteq P^n$. Since R is gr-multiplication, there exists a graded ideal B such that $A = BP^n$. If $B \subseteq P$, then $A \subseteq P^{n+1}$ and therefore, $A = P^{n+1}$. If not, $B + P = R$ and hence, $P^n = P^n(B + P) = P^{n+1} + P^nB = A$. ∎

The next result is a local characterization of gr-multiplication rings similar to the achieved one for the ungraded case.

Theorem 4.4.5 *A gr-noetherian ring R graded by an ordered group is gr-multiplication if and only if R^g_P is a gr-SPIR for every $P \in Spec^{gr}(R)$.*

Proof: Firstly, we shall prove that every gr-local gr-multiplication gr-noetherian ring graded by an ordered group is a gr-SPIR. Let R be a gr-local gr-multiplication ring. If R is a domain, then it is a gr-Dedekind domain by Proposition 4.4.2. Now, [91, Theorem II.2.1], implies that R is a gr-DVD and, therefore, a gr-SPIR. Now, suppose that R is not a domain. By Proposition 4.2.17, the only gr-maximal ideal, P, can be written as $P = (x)$ for some $x \in h(R)$. Since R is not a domain and G is an ordered group, by [91, Proposition A.II.1.4], there exist some homogenous zero divisors in R. Now, it will be proved that if $a \in h(R)$ is a zero divisor, then there exist a unit $u \in R$ and a positive integer m such that $a = ux^m$. In fact, since $(a) \in \mathcal{L}^{gr}(R)$, there exists a non zero element $a_1 \in h(R)$ such that $a = a_1x$. If a is a unit, the process ends. If not, $a_1 \in (x)$ and therefore, there exists another non zero homogeneous element a_2 verifying $a_1 = a_2x$ and $a = a_2x^2$. As before, if a_2 is a unit, we have finished. If not, we repeat the process. If the wished property were not true, we would obtain an ascendent chain of graded ideals (as their generators are homogeneous), $(a_1) \subseteq (a_2) \subseteq \cdots \subseteq (a_n) \subseteq \cdots$. Since R is gr-noetherian, there exists n such that $(a_n) = (a_{n+1})$. Then, $a = a_nx^n = a_{n+1}x^{n+1}$ with $a_n = a_{n+1}x$. It holds that $a_{n+1} = a_n.b$ for some $b \in h(R)$ as $a_{n+1} \in (a_n)$. It is deduced that $a_{n+1}(1 - bx) = 0$, but $1 - bx$ is a unit because it is homogeneous and it does not belong to P. It follows that $a_{n+1} = 0$ and , consequently, $a = 0$, which is a contradiction. Thus there exist zero divisors, $a = ux^n$ and $b = vx^m$, $ab = 0$ and u, v being units. It follows that $x^{n+m} = 0$. Therefore, $P^z = 0$ for some positive integer z. By Lemma 4.3.4, $\mathcal{L}^{gr}(R)$ is totally ordered under inclusion. It suffices to apply Lemma 4.4.4 to finish the first part of the proof. By Proposition 4.4.3, the converse is clear. ■

Recall that a ring is said to be a *gr-PIR* is every graded ideal is principal. A graded ideal which is generated by a homogeneous element is called *gr-*

cyclic. In the general graded local case, we obtain the next result.

Proposition 4.4.6 *Every gr-local gr-multiplication ring is gr-PIR.*

Proof: Let R be a gr-local ring. Then, we shall prove that every gr-multiplication ideal is gr-cyclic. Let P be the gr-maximal ideal of R. Suppose that A is a non-zero proper graded ideal of R. Since A is gr-multiplication, given $0 \neq a \in h(A)$, $aR = AB_a$ for some $B_a \leq_{gr} R$. Then, $A = \sum_{a \in A} aR = \sum_{a \in A} AB_a = A(\sum_{a \in A} B_a)$. If $\sum_{a \in A} B_a = R$, then $B_a = R$ for some $a \in A$ because R is gr-local. In this case, $A = AB_a = aR$ and hence, A is gr-cyclic. If $\sum_{a \in A} B_a \neq R$, then $A = AP$. For $0 \neq a \in h(A)$, it holds $aR = AB_a = APB_a = AB_aP = aP$. By Proposition 4.2.1, $a = 0$, a contradiction. ■

From Theorem 4.1.3, we immediately obtain the following characterization of strongly gr-multiplication rings. The result can be also deduced from the category equivalence between R-gr and R_e-Mod in case R is strongly graded. We give another proof by using graded techniques.

Proposition 4.4.7 *Let R be a strongly graded ring. Then, R is gr-multiplication if and only if R_e is a multiplication ring.*

Proof: When the ring is strongly graded, R is a generator of gr-R and therefore, this category is a commutative Grothendieck category. It is well-known that, in this case, gr-R is equivalent to the category Mod-R_e. By Theorem 2.2.7 , an ideal I_e of R_e is a multiplication object in Mod-R_e if and only if I_eR is multiplication in gr-R. By Proposition 2.2.4, I_e is a multiplication object if and only if it is a multiplication ideal of R_e. By Theorem 4.1.3, I_eR is a multiplication object in gr-R if and only if it is a gr-multiplication ideal and the proof ends. ■

As a consequence of Proposition 4.4.7, we can give another family of gr-multiplication rings. Recall from [113] that the *generalized Rees ring associated to I* (an invertible ideal) is $\check{R}(I) = \Sigma_{n \in \mathbb{Z}} X^n I^n$, with the obvious $\mathbb{Z}$-graduation. It is well-known that these rings are strongly graded. By Proposition 4.4.7, if R is multiplication, then $\check{R}(I)$ is $\mathbb{Z}$-gr-multiplication. Note that this generalizes [113, Theorem 2.3].

If R is a Dedekind domain, G is a finite group and t is a generalized 2-cocycle, i. e., not necessarily taking its values in the units of R, then the generalized twisted group ring (cf. [95]), $R^t[G]$, is gr-multiplication.

The next result can be considered as a Mott's theorem (cf. [80]) for graded rings.

Theorem 4.4.8 *Let R be a ring strongly graded by an ordered group. Then, R is gr-multiplication if and only if every $P \in Spec^{gr}(R)$ is gr-multiplication.*

Proof: Suppose that every graded prime is gr-multiplication. Let Q be a prime ideal of R_e. By [91, Lemma A.I.8.15 and Lemma A.II.2.2], there exists $P \in Spec^{gr}(R)$ such that $P_e = Q$. By the assumption, P is gr-multiplication. By Theorem 4.1.3, P_e is multiplication as an R_e-module. We have proved that every prime ideal of R_e is multiplication. By Mott's theorem, R_e is multiplication. By Proposition 4.4.7, R is gr-multiplication. Since the other implication is obvious, the equivalence is proved. ■

4.5 Graded rings and multiplication rings

Now, our purpose is to determine when a graded ring is a multiplication ring. For this reason, we relate that R is gr-multiplication, that R_e is a multiplication ring and that R is multiplication.

Recall some concepts which appear in the next result. Given a ring strongly graded by G, R, and a subgroup $H \leq G$, it is well-known that the ring $R^{(H)} = \oplus_{g \in H} R_g$ is strongly graded (cf. [91, Example 1.3.1, p. 15]). Recall that a commutative ring is called *reduced* or *semiprime ring* if it has no nilpotent elements different from zero.

Proposition 4.5.1 *Let R be a ring graded by G and let H be a subgroup of G. Then,*

1. *if R is gr-multiplication ring, then R_e is multiplication.*

2. *if R is strongly graded and R is multiplication, so is $R^{(H)}$.*

3. *if R is noetherian and strongly graded and G is finite with order invertible in the reduced ring R_e, then R is multiplication if and only if R_e is multiplication.*

Proof: In order to prove 1, let A_e, B_e be ideals of R_e with $A_e \subseteq B_e$. Consider the graded ideals, $A = RA_e$ and $B = RB_e$. By hypothesis, $A = BI$ for some graded ideal I of R. In particular, their components of degree e have to be equal. Thus $A_e = \sum_{g \in G} B_g I_{g^{-1}} = B_e(\sum_{g \in G} R_g I_{g^{-1}})$ and therefore, the ideal B_e is multiplication and hence, R_e is a multiplication ring.

Now, we shall prove 2. Every graded ideal of $R^{(H)}$ is generated by its components of degree e as it is strongly graded. Suppose that $R^{(H)} J_e \subseteq R^{(H)} I_e$ with I_e and J_e ideals of R_e. It follows that $J_e \subseteq I_e$ and therefore, $RJ_e \subseteq RI_e$. By hypothesis, $RJ_e = RA_e I_e$ for some ideal A_e of R_e. By taking direct sum in H, it remains $R^{(H)} J_e = R^{(H)} A_e I_e = (R^{(H)} A_e)(R^{(H)} I_e)$.

To prove 3, suppose that R_e is multiplication. By [91, Exercise C.I.6 (3)], R is reduced. By [57, Corollary 2], R_e is hereditary. By [92, Corollary 2.12], so is R and, hence, it is multiplication. The converse holds by 1. ∎

Remarks. *i)* The implication which appears in Proposition 4.5.1 (1) is strict. Consider the polynomial ring $R = \mathbb{Z}[X]$ graded by $\mathbb{Z}$ in the usual way. By [91, Remark B.II.2.9], it is a domain but it is not a gr-Dedekind domain. By Proposition 4.4.2, R is not gr-multiplication. However, its component of degree zero, $\mathbb{Z}$, is clearly multiplication.

ii) Proposition 4.5.1 (3) gives us that the above defined generalized twisted group ring $R^t[G]$ is multiplication when the order of G is invertible in R.

Some results have been obtained in the finite case. In general, it is not always true that if R is gr-multiplication, then it is multiplication. For example, the group ring $R = \mathbb{Z}_6[\mathbb{Z}_2 \oplus \mathbb{Z}_3]$ is gr-multiplication because so is $\mathbb{Z}_6$. However, [52, Theorem 19.10] gives us that R is not multiplication.

When the order of the group G is not invertible in R, the situation is more complicated. First, recall the construction of twisted group rings. Let $\Phi : G \to Aut(R)$ be a morphism from the group G to that of automorphisms of a ring R. Denote $\Phi(g) = \alpha_g$. A *2-cocycle* is a mapping $t : G \times G \to UZ(R)$ (group of units of the center of R) verifying $t(g,h)t(gh,l) = t(h,l)^{\alpha_g} t(g,hl)$, where $t(h,l)^{\alpha_g}$ is $t(h,l)$ in the commutative case (cf. [91, p. 26]). In these circumstances, the free R-module with base G and whose product is defined as follows: $(\sum_{g \in G} a_g g)(\sum_{h \in G} b_h h) = \sum_{l \in G} c_l l$, where $c_l = \sum_{gh=l} a_g b_h^{\alpha_g} t(g,h)$, is called *crossed product of the ring R and the group G*. It will be denoted by $R * G$ and it is graded in the following way: $(R * G)_g = \{rg; r \in R\}$. If the action is trivial, then the consequential concept is that of *twisted group ring with respect to* t, which will be denoted by $R^t[G]$ (these concepts can be found in [91]). We write G_p for the p-primary component of the finite group G. When the ground ring is a field, we have the next result.

Proposition 4.5.2 *Let $K^t[G]$ be a commutative twisted group algebra, K being a field and G being a finite group. If $char(K) = 0$ or $char(K) = p$ and G_p is cyclic, then $K^t[G]$ is multiplication. Moreover, if K is perfect the converse is also true.*

Proof:

Suppose that $char(K) = 0$. Then $R = K^t[G]$ is multiplication from Proposition 4.5.1 (3). Let $p = char(K) > 0$. If p does not divide the order of G, then by Maschke's Theorem, $K^t[G]$ is semisimple and therefore, it is multiplication. Suppose that p divides the order of G and G_p is cyclic. By [65, Theorem III.2.8], R is isomorphic to a quotient of a group algebra $K[H]$ where H has the same p-primary component as G. By [52, Theorem 19.10], $K[H]$ is multiplication and, therefore, R is multiplication as well.

Conversely, suppose that $K^t[G]$ is multiplication and $char(K) = p > 0$, K being a perfect field. By Proposition 4.5.1 (2), $K^t[G_p]$ is also multiplication. Since K is perfect, we can apply [65, Lemma V.8.13] to obtain $K^t[G_p] \cong K[G_p]$. By [52, Theorem 19.10], G_p is cyclic. ■

Remarks. *i)* If K is not perfect, then the result is not always true. In fact, it suffices to consider $R = \mathbb{Z}_2(X)^t[\mathbb{Z}_2 \oplus \mathbb{Z}_2]$ where $\mathbb{Z}_2(X)$ is the field of rational functions with coefficients in $\mathbb{Z}_2$ and t verifies that its restriction to $\mathbb{Z}_2$ is not coboundary. Then, $\mathbb{Z}_2(X)^t[\mathbb{Z}_2]$ is a field (see [65, p. 253]) and, by the first part of Proposition 4.5.2, R is multiplication (the restriction of t has been denoted by t as well).

ii) There are examples of twisted group rings and therefore, strongly graded by a finite group, which are multiplication or not depending on the cocycle. For instance, the twisted group ring $\mathbb{Z}_8^t[C_4]$ where the 2-cocycle t is defined as follows: $t(0,0), t(1,0), t(1,1), t(3,0), t(2,0)$ and $t(2,1)$ are equal to 1, and $t(3,3) = 3 = t(3,2) = t(3,1) = t(2,2)$. Of course, t is symmetrical. This ring is clearly isomorphic to $\mathbb{Z}_8[X]/(X^4 - 3)$, which is a SPIR (see [28]) and, therefore, it is a multiplication ring. However, if we consider the graded ring $R = \mathbb{Z}_8^t[C_4]$ where now, the cocycle t is the convenient one to establish the isomorphism with $\mathbb{Z}_8[X]/(X^4-5)$, then R is not multiplication. Accordingly, it seems interesting to study the influence of the cocycle in order to obtain a multiplication ring and we will deal with this problem in a subsequent study. Henceforth our work will be focussed on graded ring whose grading group is torsion-free.

First, the local case will be studied.

Lemma 4.5.3 *If R is a local ring strongly graded by $\mathbb{Z}$, then R is a group ring.*

Proof: Firstly, it will be proved that if R is a graded ring which is local, then R_e is local and therefore, R is gr-local. Assume that P is the only maximal ideal of R and that there is two different maximal ideals P_1 and P_2 of R_e. Then, $P_1 + P_2 = R_e$. The graded ideals RP_1 and RP_2 are different because so are their components of degree e and they are proper ideals because their components of degree e are different from R_e. Thus, $RP_1, RP_2 \subseteq P$. It holds

$R_e = P_1 + P_2 \subseteq P \cap R_e$. Hence, $R = RR_e \subseteq P$, which is a contradiction. The equivalence between that R_e is local and that R_e is gr-local give it to us [16, Proposition 2].

Since R is local, R_e is local. By [91, Corollary A.I.3.26], R is a crossed product. By [91, Corollary A.I.3.24], R is a group ring, with ground ring R_e. ■

If R is a domain strongly graded by $\mathbb{Z}$, it is obtained the following structure theorem.

Theorem 4.5.4 *Let R be an integral domain strongly graded by $\mathbb{Z}$. Then, R is multiplication if and only if R is a group ring whose ground ring is a field, i. e., $R \cong K[X, X^{-1}]$, K being a field.*

Proof: If R is a multiplication domain, then it is a Dedekind domain and, in particular, it is a Prüfer domain. By [29, Lemma 4.2], $R \cong K[\mathbb{Z}]$, where K is a field. It follows from [52, Theorem 13.8] that $K[\mathbb{Z}]$ is a Dedekind domain. ■

The next result extends the case of finite direct sums of P.I.D.'s and S.P.I.R.'s which appears in [28, Proposition 1] and it is necessary to prove the following structure theorem.

Lemma 4.5.5 *Let R be a noetherian multiplication ring (strongly) graded by a torsion-free group G. Then, R is finite direct sum of Dedekind domains and special primary rings, each of them being (strongly) graded by G.*

Theorem 4.5.6 *Let R be a noetherian ring strongly graded by $\mathbb{Z}$. Then, R is multiplication if and only if R_e is a finite product of fields. Moreover, in this case, R is a group ring.*

Proof: First, suppose that R is multiplication. Then,

$$R = D_1 \oplus \cdots \oplus D_r \oplus S_1 \oplus \cdots \oplus S_t,$$

where D_i $(1 \leq i \leq r)$ is a Dedekind domain and S_j $(1 \leq j \leq t)$ is a special primary ring (SPIR). By [28, Proposition 1], so are the D_i's and the S_j's are strongly graded by G. The degree e component of R is

$$R_e = (D_1)_e \oplus \cdots (D_r)_e \oplus (S_1)_e \oplus \cdots \oplus (S_t)_e.$$

By Proposition 4.5.4, $(D_i)_e$ is a field for $1 \leq i \leq r$. If S is a SPIR, it is local. If, moreover, it is strongly graded, by Lemma 4.5.3, it is group ring. However, $S[\mathbb{Z}]$ cannot be local by [52, Corollary 19.2]. It follows that R_e is a finite direct sum of fields.

Now, suppose that R_e is a finite direct product of fields. By [91, Corollary I.3.26], R is a crossed product of R_e and, by [91, Corollary I.3.24], $R \cong R_e[\mathbb{Z}]$. By [52, Theorem 18.10], it is deduced that R is multiplication. ∎

In the following case, an equivalence between being gr-multiplication and being multiplication is achieved.

Theorem 4.5.7 *If R is a gr-noetherian ring positively graded by $\mathbb{Z}$, then R is gr-multiplication if and only if it is multiplication.*

Proof: In the first place, we claim that if R is gr-noetherian and gr-multiplication (for any group G), then every non-zero graded ideal is a finite product of graded prime ideals. Let $A \leq_{gr} R$. If $A \in Spec^{gr}(R)$, we have finished. If not, there exists $P \in Max^{gr}(R)$ such that $A \subseteq P$. Since R is gr-multiplication, $A = PB$, where $B = (A : P)$, which is a graded ideal of R containing A. If $B \in Spec^{gr}(R)$, the process finishes. If not, we can repeat the above reasoning. It is obtained an ascendent chain of graded ideals $A \subseteq B \subseteq C \cdots$. Because R is gr-noetherian, the process ends and therefore, A can be written as a finite product of graded prime ideals.

Suppose that R is gr-multiplication. By the first part, every non-zero graded ideal of R is a product of graded prime ideals. By [9, Theorem 7], every non-zero ideal (graded or not) is a finite product of prime ideals. By [69, Theorem 9.10], R is a finite direct sum of Dedekind domains and SPIR's and hence, it is multiplication. The other implication is always true. ■

If the ring is graded (not necessarily strongly graded) by a torsion-free group, noetherian multiplication rings can also be characterized in the following way.

Theorem 4.5.8 *Every noetherian multiplication ring graded by a torsionfree group, is a finite direct sum of trivially graded Dedekind domains, trivially graded SPIR's and rings of the form $K[X]$, $K'[\mathbb{Z}]$ and rings of the form $K''[X]/(X^n)$ with $n \geq 2$, K, K' and K'' being fields.*

Proof: Since R is multiplication and noetherian, it is a direct sum of a finite number of Dedekind domains and SPIR's, which will be graded by the same group G. By [27, Corollary 5], domains which are not trivially graded will be of the form $K[X]$ or $K[\mathbb{Z}]$ for some field K. By [28, Proposition 3], the non-trivially graded SPIR's are of the form $K[X]/(X^n)$. It suffices to join these results to deduce the theorem. ■

Bibliography

[1] ABDULLAH, MUSTAFA and NAOUM, *On finitely generated projective modules and multiplication modules*, Periodica Math. Hungar., **20** (1), (1989) 57-63.

[2] ALBU, T. and NĂSTĂSESCU, C., *Modules arithmétiques*, Acta Math. A. S. Hung. **25**(3-4), (1974) 299-311.

[3] ALBU,T. and NĂSTĂSESCU, C., *Décompositions primaires dans les catégories de Grothendieck commutatives. I*, Journal für die reine und angewandte Mathematik, Sonderdruck aus Band **280**, (1976) 172-194.

[4] ALBU,T. and NĂSTĂSESCU, C., *Décompositions primaires dans les catégories de Grothendieck commutatives. II*, Journal für die reine und angewandte Mathematik, Sonderdruck aus Band **282**, (1976) 172-185.

[5] ALBU, T., *Une remarque sur les catégories de Grothendieck commutatives*, Bull. Math. Soc. Sci. Math. R. S. Roumanie, **23** (71), (1979) 115-116.

[6] ALBU, T., *On commutative Grothendieck categories having a Noetherian cogenerator*, Arch. Math., Vol. **34**, (1980) 210-219.

[7] ALBU, T. and NĂSTĂSESCU, C., *Relative Finiteness in Module Theory*, Monographs and Textbooks in Pure and Applied Mathematics **84**, New York (1984).

[8] ANDERSON, D.D., *Multiplication Ideals, Multiplication Rings, and the Ring $R(X)$*, Can. J. Math., Vol. **XXVIII**, No. 4, (1976) 760-768.

[9] ANDERSON, D. D. and MATIJEVIC, J., *Graded π-rings*, Can. J. Math., no.3, (1979) 449-457.

[10] ANDERSON, D.D., *Some remarks on multiplication ideals*, Math. Japonica **25**, No.4, (1980) 463-469.

[11] ANDERSON, D.D., MOTT, J. and ZAFRULLAH, M., *Some Quotient Based Statements in Multiplicative Ideal Theory*, Bolletino U.M.I., (7) 3-B (1989), 455-476.

[12] ANDERSON, F.W. and FULLER, K.R., *Rings and Categories of Modules*, Springer, New York, (1974).

[13] BARNARD, A., *Multiplication Modules*, Journal of Algebra **71**, (1981) 174-178.

[14] BARBOUL, M. and NAOUM, A.G., *On finitely generated multiplication ideals in commutative rings*, Beiträge zur Algebra and Geometrie, **19**, (1985) 75-82.

[15] BEATTIE, M. and ORZECH, M., *Prime Ideals and Finiteness Conditions on Gabriel Topologies over Commutative Rings*, Rocky Mountain J. Math. **22**, no. 2, (1992) 423-439.

[16] BEATTIE, M., DĂSCĂLESCU, S. and NĂSTĂSESCU, C., *A note on semilocal graded rings*, Rev. Roumaine Math. Pures Appl., **40** 3-4, (1995) 253-258.

[17] BÉNABOU, J., *Catégories avec multiplication*, C. R. Acad. Sci. Paris, Ser. I Math., **256**, (1963) 1887-1890.

[18] BRANDAL, W. and BARBUT, E., *Localizations of torsion theories*, Pacific J. of Math., Vol. **107**, No. 1, (1983) 27-36.

[19] BUESO J.L., TORRECILLAS B. and VERSCHOREN A., *Local cohomology and localization* Pitman Resarch Notes in Mathematics Series **226**, Longman Scientific and Technical, Harlow (1989).

[20] BUTTS, H.S. and PHILLIPS, R.C., *Almost multiplication rings*, Canad. J. Math., **17**, (1965) 267-277.

[21] CAHEN, P.J., *Torsion theory and associated primes*, Proc. Amer. Math. Soc. **18**, (1973) 471-477.

[22] CAMILLO, V., *Distributive modules*, J. of Algebra, **36**, (1975) 16-25.

[23] CAENEPEEL, S., VAN OYSTAEYEN, F. and BORONG ZHOU, *Making the category of Doi-Hopf modules into a braided monoidal category*, Algebras and Representation Theory, **1** (1), (1998) 75-96.

[24] CHARI, V. and PRESSLEY, A., *A guide to Quantum Groups*, Cambridge University Press, Cambridge, (1994).

[25] CHOUINARD, L. D., HARDY, B. R. and SHORES, T. S., *Arithmetical and Semihereditary Rings*, Comm. in Algebra, **8** (17), (1980) 1593-1652.

[26] CHEW, K.L., *Closure operations in the study of rings of quotients*, Bull. Math. So. Nanyang Univ., (1965) 1-20.

[27] DECRUYENAERE, F. and JESPERS, E., *Dedekind domains and graded rings*, Proceedings of the American Math. Soc., **110** (1), (1990) 21-26.

[28] DECRUYENAERE, F. and JESPERS, E., *Graded commutative principal ideal rings*, Bull. Soc. Math. Belg., **43** (2), Ser. B, (1991) 143-150.

[29] DECRUYENAERE, F. and JESPERS, E., *Prüfer domains and graded rings*, J. Algebra, **150**, (1992) 308-320.

[30] DICKSON, S.E., *A torsion theory for abelian categories*, Trans. Amer. Math. Soc., **121**, (1966) 233-235.

[31] EL-BAST, A. and SMITH, P. F., *Multiplication modules*, Comm. in Algebra, **16**(4), (1988) 755-779.

[32] EL-BAST, A. and SMITH, P. F., *Multiplication modules and Theorem of Mori and Mott*, Comm. in Algebra, **16** (4), (1988) 781-796.

[33] ERDOĞDU, V., *Distributive Modules*, Canad. Math. Bull., **30** (2), (1987) 248-254.

[34] ERDOĞDU, V., *Multiplication modules which are distributive*, J. Pure and Applied Algebra, **54**, (1988) 209-213.

[35] ERDOĞDU, V., *Regular Multiplication Rings*, J. of Pure and Applied Algebra, **59**, (1989) 55-59.

[36] ESCORIZA, J. and TORRECILLAS, B., *Multiplication modules relative to torsion theories.* Comm. in Algebra, **23** (11), (1995) 4315-4331.

[37] ESCORIZA, J. and TORRECILLAS, B., *Relative Multiplication and Distributive Modules*, Comment. Math. Univ. Carolinae, **38** (2), (1997) 205-221.

[38] ESCORIZA, J. and TORRECILLAS, B., *Multiplication Objects in Commutative Grothendieck Categories.* Comm. in Algebra, **26** (6), (1998) 1867-1883.

[39] ESCORIZA, J. and TORRECILLAS, B., *Multiplication graded rings*, Algebra and Number Theory, Lecture Notes in Pure and Applied Mathematics, vol. **226**, Marcel Dekker, (1999) 127-136.

[40] ESCORIZA, J. and TORRECILLAS, B., *Multiplication rings and graded rings*, Comm. in Algebra, **27** (12), (1999).

[41] ESCORIZA, J. and TORRECILLAS, B., *Multiplication objects in monoidal categories.* Preprint.

[42] ESCORIZA, J. and TORRECILLAS, B., *Relative multiplication rings.* Prepint.

[43] ESCORIZA, J. and TORRECILLAS, B., *Multiplication object in semimodules.* Preprint.

[44] FOSSUM, R.M., *The divisor class group of a Krull domain*, Springer, Berlin (1987).

[45] FUCHS, L., *Über die Ideale arithmetischer Ringe*, Comment. Math. Hel., **23**, (1965), 334-341.

[46] GABRIEL, P., *Des catégories abéliennes*, Bull. Soc. Math. France, **90**, (1977) 33-46.

[47] GARCÍA, J.M., JARA, P. and SANTOS, E., *Prüfer $*$-multiplication domains and torsion theories*, Comm. in Algebra, **27** (3), (1999) 1275-1295.

[48] GARCÍA HERNÁNDEZ, J.L. and GÓMEZ PARDO, J.L., *On Endomorphism Rings of Quasiprojective Modules*, Math. Z., **196**, (1987) 87-108.

[49] GELFAND, S. and KAZHDAN, D., *Examples of tensor categories*, Invent. Math., **109**, (1992) 595-617.

[50] GILMER, R.W., Jr. and MOTT, J.L., *Multiplications rings as rings in which ideals with prime radical are primary*, Trans. Amer. Math. Soc., **114**, (1965) 40-52.

[51] GILMER, R.W., *Multiplicative ideal theory*, Marcel Dekker, New York (1972).

[52] GILMER, R.W., *Commutative Semigroup Rings*, The University of Chicago Press, Chicago (1984).

[53] GOLAN, J.S., *Torsion Theories*, Pitman Monographs and Surveys in Pure and Applied Mathematics, **29**. Longman Scientific and Technical, Harlow (1986).

[54] GOLAN, J.S., *The theory of semirings with applicatios in mathematics and theoretical computer science*, Longman Scientific and Thecnical, John Wiley and Sons, Inc., New York (1992).

[55] GÓMEZ PARDO, J.L., *Rings of quotients of endomorphism rings*, Lectures Notes in Math., Proc. Granada Ring Theory, (1986) 106-121.

[56] GOODEARL, K.R., *Ring Theory, nonsingular rings and modules*, Pure and Applied Mathematics, **33**, Marcel Dekker, New York (1976).

[57] GRIFFIN, M., *Multiplication rings via their total quotient rings*, Can. J. Math. **Vol. XXVI** (2), (1974) 430-449.

[58] HASAN, M.A.K. and NAOUM, A.G., *The residual of finitely generated multiplication modules*, Arch. Math. **46**, (1986) 225-230.

[59] JENSEN, C.U., *Arithmetical rings*, Acta Math. Acad. Sci. Hungar., **17** (1-2), (1966) 115-123.

[60] JEREMÍAS LÓPEZ, A., *El grupo fundamental relativo. Teoría de Galois and localización*, Álxebra **56**, Santiago de Compostela, 1991.

[61] JEREMÍAS LÓPEZ, A., LÓPEZ LÓPEZ, M.P. and VILLANUEVA NOVOA, E., *Gabriel filters in Grothendieck categories*, Publicacions Matemàtiques, **36**, (1992) 673-683.

[62] JEREMÍAS LÓPEZ, A., LÓPEZ LÓPEZ, M.P. and VILLANUEVA NOVOA, E., *Localization in symmetric closed Grothendieck categories*, Bull. Soc. Math. de Belgique, **45**, (1993) 198-221.

[63] JESPERS, E., KREMPA, S. and PUCZYLOWSKI, E.R., *On radicals of graded rings*, Comm. in Algebra, **10** (17), (1982) 1849-1854.

[64] JOYAL, A., *Braided Tensor Categories*, Advances in Math., **102**, (1993) 20-78.

[65] KARPILOVSKY, G., *Projective representations of finite groups*, Marcel Dekker, New York (1985).

[66] KASSEL, C., *Quantum groups*, Graduate texts in Mathematics, **155**, Springer, Berlin (1995).

[67] KIRBY, D., *Closure operations on ideals and submodules*, J. London Math. Soc., **44**, (1969) 283-291.

[68] KRULL, W., *Über Multiplicationsringe*, S.-B. Heidelberger Akad. Wiss, **5**, (1925) 13-18.

[69] LARSEN, M. D. and McCARTHY P.J., *Multiplicative theory of ideals*, Pure and Applied Mathematics, **43**, Academic Press, New York (1971).

[70] LE BRUYN, L., VAN DER BERGH and VAN OYSTAEYEN, F., *Graded Orders*, Birkhauser, Boston (1988).

[71] LÓPEZ LÓPEZ, M.A. and VILLANUEVA NOVOA, E., *The Brauer Group of the Category* (R,σ)-**mod** *(An Alternative to the Theory of Relative Invariants)*, Proc. of the First Belgian-Spanish Week on Algebra and Geometry, (1988).

[72] LOW, G. M. and SMITH, P. F., *Multiplication Modules and Ideals*, Comm. in Algebra, **18** (12), (1990) 4353-4375.

[73] MAC LANE, S., *Natural associativity and commutativity.* Rice Univ. Studies, **49**, (1963) 28-46.

[74] MAC LANE, S., *Categories for the working mathematician*, Graduate Texts in Mathematics **5**, Springer, Berlin (1971).

[75] MEHDI, F., *On Multiplication Modules*, Mathematics Student, Vol. **XLII**, No. 2, (1974) 149-153.

[76] MEHDI, F., *On multiplication and weak multiplication modules*, Riv. Mat. Univ. Parma **2** (4), (1976) 107-112.

[77] MEHDI, F. and SINGH, S., *Multiplication modules*, Canad. Math. Bull. Vol **22** (1), (1979) 93-98.

[78] MENINI, C. and NĂSTĂSESCU, C., *When is R-gr equivalent to the category of modules?*, J. Pure and Applied Algebra, **51**, (1988) 277-291.

[79] MONTGOMERY, S., *Hopf algebras and their actions on rings*, CBMS **82**, Amer. Math. Soc., Providence, (1993).

[80] MOTT, J.L., *Equivalent Conditions for a ring to be a multiplication ring*, Can. J. Math., **16**, (1964) 429-434.

[81] MOTT, J.L., *Multiplication Rings Containing Only Finitely Many Minimal Prime Ideals*, J. Sci. Hiroshima Univ. Ser. A-I, **33**, (1969) 73-83.

[82] NAOUM, A. G., AL-HASHIMI, B. and SHARAF, K. R., *The dual of a Finitely Generated Multiplication Module*, Periodica Math. Hungar., Vol. **20** (1), (1989) 65-74.

[83] NAOUM, A. G., *On the ring of endomorphisms of finitely generated multiplication modules*, Periodica Mathematica Hungarica, Vol. **21** (3), (1990) 249-255.

[84] NAOUM, A. G., AL-HASHIMI, B. and KIDER, J.R., *On the module of homomorphisms on finitely generated multiplication modules I*, Periodica Math. Hugar., Vol. **22** (2), (1991) 97-105.

[85] NAOUM, A. G., AL-HASHIMI, B. and KIDER, J.R., *On the module of homomorphisms on finitely generated multiplication modules II*, Periodica Math. Hugar., Vol. **22** (3), (1991) 153-160.

[86] NAOUM, A.G., *A note on projective modules and multiplication modules*, Beiträge zur Algebra and Geometrie, **32**, (1991) 27-32.

[87] NAOUM, A. G. and AL-AUBAIDY, K. H., *A note on multiplication modules and their rings of endomorphisms*, Kyugpook Math. J., **35**, (1995) 223-228.

[88] NAOUM, A. G. and AL-ALWAN, F. H., *Dedekind modules*, Comm. in Algebra, **24** (2), (1996) 397-412.

[89] NAOUM, A. G. and AL-ALWAN, F. H., *Dense submodules of multiplication modules*, Comm. in Algebra, **24** (2), (1996) 413-424.

[90] NĂSTĂSESCU, C., *La structure des modules par rapport à une topologie additive*, Tôhoku Math. Journal, **26**, (1974) 173-201.

[91] NĂSTĂSESCU, C. and VAN OYSTAEYEN, F., *Graded rings theory*, North-Holland, Amsterdam (1982).

[92] NĂSTĂSESCU, C., *Strongly graded rings of finite groups*, Comm. in Algebra, **11** (10), (1983) 1033-1071.

[93] NĂSTĂSESCU, C., *Group rings of graded rings. Applications*, J. of Pure and Appl. Math., **33**, (1984) 313-335.

[94] NĂSTĂSESCU, C. and VAN OYSTAEYEN, F., *Dimensions of Ring Theory*, Mathematics and its applications, D. Reidel Publishing Company, Dordrecht (Holland), (1987).

[95] NAUWELAERTS, E. and VAN OYSTAEYEN, F., *Generalized twisted group rings of finite groups.* Prepint.

[96] PAKALA, J. V. , *Commutative Torsion Theories*, Ph. D. thesis, University of Nebraska (1979).

[97] PANAITE, F. and STEFAN, D., *When is the category of comodules a braided tensor category?*, Rev. Roumaine Math. Pures Appl., **42** (1-2), (1997) 107-119.

[98] PIRTLE, E.M., *Integral domains which are almost Krull*, J. Sci. Hiroshima Univ., **32** (1968), 441-447.

[99] ROTMAN, J., *An introduction to homological algebra*, Pure and Applied Mathematics, Academic press, New York (1979).

[100] ROWEN, L., *Ring Theory. Volume I*, Pure and Applied Mathematics, **127**, Academic Press, San Diego (1988).

[101] SMITH. P.F., *On Non-commutative AM-rings*, Houston J. of Math., Vol. **11** (3), (1985) 405-422.

[102] SMITH, P.F., *Some remarks on multiplication modules*, Arch. Math. **50**, (1988) 223-235.

[103] SMITH, P. F., *Multiplication Modules and Projective Modules*, Peridica Math. Hungar., Vol. **29** (2), (1994) 163-168.

[104] SMITH, W. W., *Projective ideals of finite type*, Canad. J. Math., **21**, (1969) 1057-1061.

[105] STENSTRÖM, B., *Rings of Quotients*, Springer, Berlin (1975).

[106] STEPHENSON, W., *Modules whose lattice of submodules is distributive*, Proc. London Math. Soc., **3** No. 28, (1974) 291-310.

[107] SWEEDLER, M.E., *Hopf Algebras*, W. A. Benjamin, Inc., New York (1969).

[108] UKEGAWA, T., *Some Properties of Non-Commutative Multiplication Rings*, Proc. Japan Acad., **54** (9), Ser. A, (1978) 279-284.

[109] UKEGAWA. T., *Left noetherian multiplication rings*, Osaka J. Math., **17**, (1980) 449-453.

[110] UKEGAWA, T., *Some remarks on M-rings*, Math. Japonica, **28** No. 2, (1983) 195-203.

[111] VÁMOS, P., *Finitely generated artinian and distributive modules are cyclic*, Bull. London Math. Soc., **10**, (1978) 287-288.

[112] VAN OYSTAEYEN, F. and VAN DEUREN, J. P., *Arithmetically graded rings. I*, Lectures Notes in Math., **825**, Ring Theory (Antwerp 1980), Proceedings, Springer, Berlin, (1980).

[113] VAN OYSTAEYEN, F., *Generalized Rees rings and arithmetical graded rings*, J. Algebra, **82**, (1983) 185-193.

[114] VAN OYSTAEYEN, F. and VERSCHOREN, A., *Relative invariant of rings. The commutative theory.*, Lecture Notes in Pure and Applied Math., **79**, M. Dekker, New York (1983).

[115] VAN OYSTAEYEN, F., *Note on graded von Neumann regular rings*, Rev. Roumaine Math. Pures Appl., **29** (3), (1984) 263-271.

[116] WALKER C.L. and WALKER, E.A., *Quotient categories and rings of quotients*, Rocky Mountain J. Math., **2**, (1972) 513-555.

[117] WISBAUER, R., *Foundations of Module and Ring Theory*, Gordon and Brench Publishers. Philadelphia (1991).

[118] YETTER, D.N., *Quantum groups and representations of monoidal categories*, Math. Proc. Camb. Phil. Soc., **108**, (1990) 261-290.

Index

annihilator, 154
axiom
pentagon, 41
triangle, 41
biself-generator, 54
category
braided, 72
commutative Grothendieck, 43
Grothendieck, 43
monoidal, 40
monoidal equivalent, 42
strict braided, 73
strict monoidal, 41
closure
Kirby, 31
lattice, 28
coalgebra
cosemisimple, 69
multiplication, 69
comodule
cyclic, 69
multiplication, 69
cotensor product, 68
crossed product, 172
domain
$*$-multiplication, 139
almost Dedekind, 101, 142
completely integrally closed, 86
140
discrete graded-valuation, 166
graded Dedekind, 165
element
almost integral, 86, 140
homogeneous, 146
multiplication, 82
frame, 14
functor
monoidal, 41
Gabriel filter, 27

Gabriel topology, 27

ideal

τ-critical, 33

τ-invertible, 37

divisorial, 140

fractional, 86

gr-maximal, 147

graded prime, 147

radical of an, 30

star-ideal, 139

v-ideal, 140

module

M-faithful, 59

M-generated, 57

M-projective, 59

P-torsion, 102

τ-Bezout, 123

τ-cocritical, 33

τ-cyclic, 90

τ-distributive, 118

τ-multiplication, 61

τ-projective, 97

τ-quasiprojective, 101

τ-simple, 87

τ-torsion, 26

τ-torsionfree, 26

τ-totally invariant, 93

τ-uniserial, 119

balanced, 54

Dedekind, 14

distributive, 118

finitely cogenerated, 157

finitely projective, 12

gr-cyclic, 160

gr-distributive, 160

gr-multiplication, 66, 148

graded P-cyclic, 151

graded P-torsion, 151

invertible, 57

multiplication, 55

of quotients, 37, 60

semi-multiplication, 14

strongly τ-multiplication, 84

strongly graded, 146

suspension of a graded, 147

weak multiplication, 14

object

indecomposable, 76

multiplication, 48, 53

quotient, 50

simple, 49
totally invariant, 73
operation
componental, 31
star-operation, 138
v-operation, 140

quasi-bialgebra, 67

radical
graded Jacobson, 147
ring
τ-distributive, 118
τ-noetherian, 29
divisorially multiplication, 142
finitely cogenerated, 157
generalized Rees, 170
gr-local, 147
gr-multiplication, 165
gr-noetherian, 147
gr-PIR, 168
gr-regular, 166
graded, 146
graded special primary, 166
Morita equivalent, 55
of quotients, 37
reduced, 171
regular multiplication, 10
semicentered, 33, 34
special primary, 100
strongly graded, 146
twisted group, 173
ZPI-ring, 8

self-generator, 54
semi-coalgebra, 67
semimodule, 81
multiplication, 82
semiring, 81
submodule
S-saturated, 35
τ-closed, 28, 61
τ-closure of, 27
τ-torsion, 61
gr-inessential, 162
prime, 12
subgenerated, 57

torsion theory, 26
finite type, 35
having sufficient τ-criticals, 34
perfect, 37
semicentered, 34
two-cocycle, 172

unit. 40